Observing Nature by
Canoe & Kayak

Michael Runtz

Firefly Books

A FIREFLY BOOK

Published by Firefly Books Ltd. 2024
Copyright © 2024 Firefly Books Ltd.
Text and photographs copyright © 2024 Michael Runtz

All rights reserved. No part of this publication may be reproduced, stored in a retrieval system, or transmitted in any form or by any means, electronic, mechanical, photocopying, recording or otherwise, without the prior written permission of the Publisher.

First printing

Library of Congress Control Number: 2024930625

Library and Archives Canada Cataloguing in Publication
Title: Observing nature by canoe & kayak / Michael Runtz.
Other titles: Observing nature by canoe and kayak
Names: Runtz, Michael, author, photographer.
Description: Includes index.
Identifiers: Canadiana 20240288262 | ISBN 9780228104681 (softcover)
Subjects: LCSH: Canoes and canoeing. | LCSH: Kayaking. | LCSH: Nature study. | LCSH: Stream ecology—Pictorial works. | LCSH: Forest ecology—Pictorial works. | LCGFT: Illustrated works.
Classification: LCC GV783 .R86 2024 | DDC 797.122—dc23

Published in the United States by	Published in Canada by
Firefly Books (U.S.) Inc.	Firefly Books Ltd.
P.O. Box 1338, Ellicott Station	50 Staples Avenue, Unit 1
Buffalo, New York 14205	Richmond Hill, Ontario L4B 0A7

Cover and interior design: Hartley Millson

Printed in China | E

We acknowledge the financial support of the Government of Canada.

Contents

Birds	7
Mammals	85
Herptiles	101
Insects	117
Other Animals	177
Signs of Animals	185
Wildflowers	201
Woody Plants	257
Other Plants	315

Nature by Canoe and Kayak

There is no finer experience than canoeing or kayaking at the break of dawn. After a cool night, ghostly mists dance from the water, mesmerizing you until the wails of loons stir your soul, temporarily breaking the spell. When the sun finally warms its way through the haze and touches your face, the chill in your bones evaporates with the mist. From nearby forests, bird song starts with hesitation, then gains momentum and soon erupts into a chorus. As you drift in solitude, the world around you comes alive. You are one with Nature.

Whether it is along a narrow creek, winding river, or endless lake shore, traveling by canoe or kayak provides myriad encounters with Nature. Some, such as a Great Blue Heron rising on colossal wings or a magnificent Moose lifting its head from the water, aquatic plants dripping from its mouth, are recognizable sightings. Many other encounters, however, pass with much less fanfare or go completely unnoticed.

It is my hope that this book provides a name to many of the plants and animals you meet while paddling, as well as introduces you to others you may have previously overlooked. Be aware, though, that closer scrutiny of the plants and animals around you will most certainly add more time to your travels. If it does, I am sure you will find that time very well spent indeed!

How to Use this book

It would take a library of books to cover all the species one could potentially see from a canoe or kayak in Northeastern waterways. Yet, while understandably not comprehensive, this book highlights more than 350 species of plants and animals commonly encountered in that region. In a few cases, where a number of species look very much alike and require close scrutiny to identify to species level (i.e. Pondweeds), identification only to their group level is given. If the species you are looking at looks similar to but not exactly the same as one illustrated in the following pages, check out one of the excellent reference books listed in the Appendix for other options.

The book is separated into sections by taxa (Mammals, Birds, etc.) with most large groups divided into subgroups. First, go to the section to which your plant or animal likely belongs. Then look at the subgroups in that category.

Birds

Birds of the Water includes birds that swim, dive and feed in the water. Birds of the Marshes contains birds that inhabit cattail marshes and other vegetated wetlands, while Birds of the Air includes birds often seen flying or foraging over waterways. Birds of the Shoreline are birds encountered on the shore or in the trees next to it.

BIRDS | OF THE WATER

Canada Goose and Cackling Goose

1 With its white "chin strap" and distinctive honking, a Canada Goose is one of our most identifiable birds. Both sexes take care of their young — a feature not characteristic of ducks.

2 Look in large flocks of migrating Canada Geese for a Cackling Goose, a pint-sized version of a Canada Goose with a stubby bill and short, thick neck.

OF THE WATER | **BIRDS**

Mute Swan

Beautiful and native to Eurasia, Mute Swans are now well-established birds in northeastern marshes. They are aggressive and don't tolerate other species nesting near them.

BIRDS | OF THE WATER

Trumpeter Swan

Although native to western North America, Trumpeter Swans were mistakenly released in the Northeast where they are currently nesting and expanding their range. Tundra Swans, which are native and nest in the Far North, look quite similar but usually have a yellow spot in front of their eyes. Also, a Trumpeter's bill seems more massive with a flatter, straighter profile.

Tundra Swan

You will only come across a Tundra Swan during their migration. At a distance, it is difficult to tell Tundra from Trumpeter swans. However, many Tundra Swans sport a yellow spot in front of their eyes. Also, the black from the bill that reaches a Tundra's eye is much narrower than that of a Trumpeter, whose bills appear more massive with a straighter, sloped profile.

Wood Duck

Male Wood Ducks are beautiful birds. Similar to other small ducks, Wood Ducks nest in tree cavities, often some distance from the water. The sexes have different appearances and different vocalizations: females give a distinctive "ooo-eek" while males utter a raspy, rising "jeee-eet."

Blue-winged Teal

1 On the water, a male Blue-winged Teal's white facial crescent is distinctive, while a female is just a small plain duck.

2 In flight, the name-giving blue in the wings makes this species easy to identify — that is unless you encounter its larger western relative, the Northern Shoveler, which displays a nearly identical wing pattern.

Mallard

With its bright green head, a male Mallard is distinctive, but females look like American Black Ducks. However, in comparison, female Mallards are lighter brown, their bill has some orange coloration, and their wing speculum is blue and bordered with white.

American Black Duck

American Black Ducks nest farther north than Mallards, with which they hybridize in overlapping ranges. Both sexes are quite dark, and their colorful wing patch, the speculum, is purple and lacks white borders.

BIRDS | OF THE WATER

Green-winged Teal

Teal are small ducks and fast fliers, and Green-winged Teal, true to their name, sport an iridescent green wing speculum.

Ring-necked Duck

Although a male Ring-necked Duck has a brown ring on its neck, it is difficult to see. A much better name for this diving duck would be "Ring-billed Duck," because this field mark is highly visible on both sexes.

BIRDS | OF THE WATER

Common Goldeneye

Named for their brilliant eye color, Common Goldeneyes are boreal forest nesting ducks. If you are paddling in July, when these small ducks nest, you will likely encounter only females, as by then the males have flown north to molt.

OF THE WATER | **BIRDS**

Hooded Merganser

Mergansers are diving ducks that have slender bills equipped with tooth-like serrations for holding aquatic prey. Hooded Mergansers, the smallest members of their group, eat small fish, but also capture other aquatic prey, such as dragonfly nymphs and crayfish.

BIRDS | OF THE WATER

Common Merganser

1 A trait of diving birds, the legs of Common Mergansers are placed at the rear of their bodies to provide better propulsion underwater. This positioning, however, forces them to run across the water to get airborne.

2 After mating, males head north to molt, leaving incubation and other parental duties to the females. Up north, they form large all-male groups that spend much of their time fishing.

OF THE WATER | BIRDS

Pied-billed Grebe

You're more likely to hear the loud "caow caow caow" song of a Pied-billed Grebe than spot the singer. These tiny, lobe-toed, diving marsh birds can sink below the surface by expelling air from between their feathers and internal air sacs.

BIRDS | OF THE WATER

Common Loon

The wild songs of Common Loons are familiar to anyone who camps by a healthy lake in the summer. Both parents take care of their young, which hitch rides on their parents' backs while still small enough.

Double-crested Cormorant

1 Double-crested Cormorants are diving, fish-eating birds that resemble loons but have a longer, more slender neck and a hooked bill. Also, when they are on the water, they tilt their heads upward. When perched, Cormorants spread their wings to presumably help dry their wet feathers.

2 Cormorants are colonial nesters whose acidic droppings kill the vegetation on the ground and even the trees in which they place their bulky stick nests.

BIRDS | OF THE WATER

Ring-billed Gull

Named after the distinctive ring on their bill, Ring-billed Gulls are colonial nesters. Smaller than Herring Gulls, they take three years to acquire their adult plumage.

Herring Gull

Boasting a near five-foot wingspan, Herring Gulls are one of the larger gulls, taking four years to acquire their adult plumage. Islands are their preferred nesting sites, and while they typically nest on the ground, they also will construct nests in trees.

BIRDS | OF THE WATER

Caspian Tern

Described by Arthur Cleveland Bent as "the king of all terns," a Caspian Tern has an impressive four-and-a-half-foot wingspan. A tern's wings and bill are more sharply pointed than those of a gull and most bear a black cap. Their calls are also different, with those of a Caspian Tern sounding prehistoric — more of a coarse, primal scream.

Common Tern and Black Tern

1 Common Terns are graceful flyers that dive head-first into the water to catch small fish. They nest in small colonies on islands or peninsulas.

2 Black Terns, like other terns, dive head-first into the water to catch fish. They nest in sheltered bays or marshes with emergent plants, often placing their eggs on floating platforms of matted vegetation.

BIRDS | OF THE WATER

Osprey

1 Ospreys are fish specialists whose reversible toe and sharp spicules on the underside of their feet help catch and hold their slippery prey. During flight, a fish is turned so that its head faces forward, making it more aerodynamic. Ospreys are commonly seen hovering over lakes before plunging in feet first to catch a meal.

2 Ospreys build massive stick nests that are often placed in dead trees near the shore.

/ BIRDS

Bald Eagle

1 It takes five years for a Bald Eagle to acquire its pure white head and tail. Younger birds are often mistaken for the much rarer Golden Eagle, which nests in the Far North and has a much smaller head.

2 Bald Eagles soar with wings held flat while Golden Eagles hold theirs in a shallow dihedral (V-shape), not unlike that of a Turkey Vulture.

BIRDS | OF THE WATER

Belted Kingfisher

Shy birds that utter loud rattling calls when disturbed, Belted Kingfishers nest in tunnels that they excavate in sand banks. Females are more brightly colored than males, having a rusty brown breast band under the upper blue-brown one owned by both sexes. Kingfishers hover before diving into the water headfirst to capture fish with their massive bills.

OF THE AIR | **BIRDS**

Common Nighthawk

Not a hawk but a relative of the Eastern Whip-poor-will, a Common Nighthawk feeds on the wing. While this species does fly at night, it often hunts in the evening, a raspy "peent" revealing its presence overhead. During migration, large numbers can be seen catching Mayflies and Caddisflies that have emerged from rivers.

BIRDS | OF THE AIR

Turkey Vulture

A Turkey Vulture's wings are held in a V-configuration (dihedral) when soaring and gliding. Be aware that its unfeathered red head is black in young birds, which could lead to confusion with the smaller Black Vulture. Turkey Vultures have one of the most developed olfactory senses of all birds, providing them with the ability to find a carcass by odor alone from as far away as two kilometers.

OF THE AIR | **BIRDS**

Red-shouldered Hawk

A loud "kee-yer, kee-yer, kee-yer" announces the presence of a Red-shouldered Hawk. These hawks often hunt along waterways, where snakes and frogs form part of their diet, which also includes chipmunks and other small mammals.

BIRDS | OF THE AIR

Broad-winged Hawk

Broad-winged Hawks often perch by the water's edge, watching for frogs and snakes. A small hawk, a Broad-winged's unique high-pitched, two-toned whistle and wide black-and-white tail bands make it easy to identify.

OF THE AIR | **BIRDS**

Red-tailed Hawk

It takes a Red-tailed Hawk two years to get its name-giving red-brown tail feathers, but its distinctive drawn-out, raspy screams (often heard in movies, even when an eagle is portrayed) are distinctive at any age. As do other hawks, Red-tails add green conifer sprigs to their nests. As these decompose, chemicals that deter parasitic insects are released.

Peregrine Falcon

During a dive to kill a shorebird, small duck or other medium-sized prey, a Peregrine Falcon can achieve speeds reaching 360 kilometers per hour! The eastern "anatum" race was almost wiped out by DDT poisoning, but has recovered due to the banning of that nasty chemical and through captive breeding programs. Today their piercing calls are commonly heard rising from the cliff faces on which they nest.

American Crow

The loud "caws" of American Crows are distinctive sounds. If you hear multiple frenzied versions of that call, look for a mob of crows harassing a hawk or large owl, especially a Great Horned.

BIRDS | OF THE AIR

Common Raven

A Common Raven can be readily told from an American Crow by its hoarse screams. Ravens have a massive beak, longer wings, and a long, wedge-shaped tail (round in crows). Additionally, Ravens commonly soar, which crows seldom do.

OF THE AIR | **BIRDS**

Bank Swallow

1 With a brown back and band across the upper breast, Bank Swallows are distinctive.

2 Bank Swallows live in colonies and excavate nesting cavities in sand banks, so look for a cluster of small holes if you paddle a river that cuts through clay or sand deposits.

BIRDS | OF THE AIR

Tree Swallow

Tree Swallows make their nests in abandoned woodpecker holes, especially those in dead trees standing in water. Sadly, Tree Swallows have become much less common in recent years.

Barn Swallow

As the name suggests, Barn Swallows commonly nest in barns, but historically they stuck their mud nests on cliff faces. In recent years, their numbers have plummeted alarmingly, as have those of many other birds that eat flying insects. Agricultural insecticides have been implicated in these declines.

BIRDS | OF THE MARSHES

Northern Harrier

Northern Harriers often hunt along the edges of waterways, flapping and gliding with long wings. Females are brown while the smaller males are gray, but both have a white patch above the base of the tail. Both harriers and owls have facial feathers forming discs that capture sound, funneling it to their ear canals.

Short-eared Owl

While most owls are nocturnal, Short-eared Owls also hunt during the day. Their flight is light and bouncy, not unlike that of a moth. As in other owls, special feathers forming facial discs collect sound and direct it to the ear canals on the sides of their head.

BIRDS | OF THE MARSHES

Virginia Rail

Virginia Rails are more commonly heard than seen. Their peculiar grunts and "kid-ick, kid-ick, kid-ick" calls rise from marshes, where their vertically compressed bodies allow them to squeeze through the cattail stalks.

OF THE MARSHES | **BIRDS**

Sora

A small member of the rail family, Soras possess a chicken-like yellow bill and utter a distinctive, whistled "whinny." While these secretive birds can be found in cattail marshes, wet sedge meadows are also favored habitats.

BIRDS | OF THE MARSHES

Common Gallinule

A Common Gallinule's long toes allow it to walk on top of soft mud and floating vegetation. Gallinules swim well but also skulk in marsh vegetation, their odd cackling cries among the more unusual sounds to arise from a wetland. These have been described as "loud, harsh, and discordant, and nearly all curiously henlike."

American Coot

American Coots are members of the rail family, but their toes are lobed instead of long and slender. Designed for swimming in vegetated waters, the lobes close when a Coot's foot is pulled forward and open when it is pushed back.

BIRDS | OF THE MARSHES

Sandhill Crane

The strident, rattling calls of a Sandhill Crane are often heard long before the bird is seen. A crane's impressive wingspan is over two meters, and it flies with stiff shallow wingbeats and soars on flat wings. Unlike Great Blue Herons, cranes fly with their necks as well as their legs outstretched.

American Bittern

When alarmed, an American Bittern points its head up and freezes, allowing its body shape and breast stripes to blend in with the vertical lines of the plants around it. The peculiar pumping sounds of males arise from this species' preferred habitat, cattail marshes.

BIRDS | OF THE MARSHES

Least Bittern

About the size and weight of a Blue Jay, Least Bitterns are minute, reclusive marsh inhabitants that, unless spotted flying briefly into view, remain well hidden among cattails. Their soft "coo-coo-coo" calls are often the only clue to their presence.

OF THE MARSHES | **BIRDS**

Great Blue Heron

1 Great Blue Herons patiently stalk shallow water in search of frogs and fish. They place their bulky stick nests high in trees, often in dead ones surrounded by water.

2 Young herons are distinct from adults due to their solid gray crown (boldly striped black and white in adults) and by being surprisingly approachable. If you paddle right up to a heron without flushing it, it is likely going to be a youngster.

BIRDS | OF THE MARSHES

Great Egret

Great Egrets were nearly hunted to extinction because of their long plumes, which were prized components of women's hats. To combat this slaughter, the precursors of the National Audubon Society were formed, which eventually chose this species as their logo.

Green Heron

Green Herons are small herons that commonly perch in trees on the water's edge. When flushed, they often utter a loud "skuewk."

BIRDS | OF THE MARSHES

Marsh Wren

Like many marsh inhabitants, Marsh Wrens are more commonly heard than seen. These tiny birds skulk in the cattails, cocking their tails over their backs when they sing. Males are polygynous and seem to try to impress potential mates by building "dummy nests" prior to their arrival.

Alder Flycatcher

The wheezy "wee-be-oh" of Alder Flycatchers is a common sound in vegetated wetlands, especially those dominated by Speckled Alders. Their call note is a single, emphatic "beeer." Alder Flyctachers tend to nest in wetter habitats than the similar-looking Willow Flycatcher.

BIRDS | OF THE MARSHES

Swamp Sparrow

Perhaps "Marsh Sparrow" would be a better name for the Swamp Sparrow because those birds are common inhabitants of cattail marshes. Their trilled song is a familiar sound to paddlers passing by that habitat.

OF THE MARSHES | **BIRDS**

Red-winged Blackbird

1 Male Red-winged Blackbirds display their bright epaulets when they sing, but at other times they can hide them under black feathers. Males have harems of females in their large territories, which they aggressively defend against intruders.

2 With their brown plumage and streaked breasts, female Red-winged Blackbirds look not unlike sparrows on steroids.

BIRDS | OF THE MARSHES

Common Grackle

Common Grackles are large blackbirds that often forage along the water's edge. Here, they are especially adept at catching dragonflies that left the water as nymphs and are transforming into their winged adult stage.

Common Yellowthroat

The "witchity-witchity-witchity" song of Common Yellowthroats rises from the thick vegetation in which these small warblers skulk. Only males have a black mask, but both sexes have the name-giving yellow throat.

BIRDS | OF THE SHORELINE

Killdeer

Killdeer reveal their identity vocally with name-giving calls. To distract predators from a nest, they put on a dramatic broken-wing act, conspicuously showing off simulated blood-stained feathers at the base of their tail as they pathetically flop on the ground.

Wilson's Snipe

Most views of a Wilson's Snipe are of a flushed bird rapidly flying in a zig-zag pattern, uttering a raspy "schkeep." In spring, males perform flight displays high in the sky with haunting laughter-like sounds arising from the outspread outer tail feathers as they vibrate in the air.

Spotted Sandpiper

Spotted Sandpipers walk along shores, constantly bobbing their back ends up and down. They fly with bursts of stiff wingbeats accompanied by distinctive "weet-weet-weet" calls. Females acquire multiple mates and saddle them with most of the parental duties, including incubation.

Solitary Sandpiper

Solitary Sandpipers nest in boreal forest trees but during migration, they typically feed alone or in small numbers along the edges of small waterways. After being flushed, a Solitary Sandpiper flies a short distance and lands, characteristically holding its wings up before slowly and gracefully lowering and tucking them away.

BIRDS | OF THE SHORELINE

Greater Yellowlegs

Greater Yellowlegs are large sandpipers that nest in the boreal forest. During migration, they feed along the edges of shallow waterways, their strident "whew-whew-whew" calls demonstrating displeasure at being disturbed by passing paddlers.

American Pipit

During spring and fall migration, flocks of American Pipits walk along shores and exposed mudflats, their tails bobbing up and down. When pipits fly, they call "sipit- sipit," and their white outer tail feathers conspicuously flash.

BIRDS | OF THE SHORELINE

Great Horned Owl

Great Horned Owls often hunt by the water's edge because Muskrats are a favored prey. The "horns" are neither horns nor ears, but decorative feathers.

OF THE SHORELINE | BIRDS

Barred Owl

Barred Owls are large owls with dark eyes that often start hunting in late afternoon. While their rhythmic "who-cooks-for-you, who-cooks-for-you-awwl" calls are given at night, it is not rare to hear them during the day.

BIRDS | OF THE SHORELINE

Yellow-bellied Sapsucker

An appropriately-named woodpecker, the Yellow-bellied Sapsucker drills neat rows of holes in many types of trees (White Birch being a favorite) to access their sap. Sapsuckers drill sap wells into a tree's xylem tissues in spring and phloem tissues in summer, keeping them open to access the sugar-rich sap they carry.

Black-backed Woodpecker

Black-backed Woodpeckers are northerners that have only three toes, a feature that might aid them in stripping bark from coniferous trees to access beetles in the wood underneath. Both sexes have a black back, but only males sport a gold crown.

BIRDS | OF THE SHORELINE

Hairy Woodpecker and Downy Woodpecker

1 A Hairy Woodpecker looks like a larger version of a Downy Woodpecker. In both species, only the males sport a red patch on the backs of their heads.

2 A Downy Woodpecker can be told from its larger look-alike by a proportionately smaller bill and black spots on its white outer tail feathers.

Northern Flicker

Only male Northern Flickers sport a black "moustache." Flickers primarily eat ants, so they spend much of their time on the ground. When a flicker flies, its conspicuous white rump and yellow wing flashes are highly visible.

BIRDS | OF THE SHORELINE

Pileated Woodpecker

1 The crow-sized Pileated Woodpecker is the largest of its group in North America. Wood-dwelling Carpenter Ants form a large part of their diet, with massive excavations resulting when a colony is discovered. These woodpeckers fly with strong, sweeping wingbeats, and are very active late in the day.

2 Males sport red "moustaches"; in females, those markings (formally known as malar stripes) are black.

OF THE SHORELINE | **BIRDS**

Eastern Kingbird

Named for their fearless harassment of any large bird that flies over their territory, Eastern Kingbirds are large flycatchers with white-tipped tails that typically place their nest on a low branch hanging over water or on stumps surrounded by it.

BIRDS | OF THE SHORELINE

Olive-sided Flycatcher

Olive-sided Flycatchers perch at the very top of tall dead trees, from which they taunt thirsty paddlers with their song: "Quick-three-beers!"

OF THE SHORELINE | **BIRDS**

Eastern Phoebe

1 Eastern Phoebes are easy flycatchers to identify, for they call their name ("fee-bee") and constantly wag their tails.

2 Phoebes often nest under bridges, but many still place their moss-covered nests just above eye-level in crevices on cliffs.

BIRDS | OF THE SHORELINE

Red-eyed Vireo and Blue-headed Vireo

1 Red-eyed Vireos, one of the most common songbirds in eastern North America, slowly work their way through vegetation, their hook-tipped bills gleaning insects from leaves. Their lazy "*cherr-o-wit, cheree, sissy-a-wit, tee-oot*" song is repeated thousands of times throughout the day.

2 Blue-headed Vireos sing with a metallic tone to their song and are the only vireos commonly encountered in coniferous habitats.

Gray Catbird

Gray Catbirds like to skulk in shrubs and other thick vegetation, their calls and songs often the only clue to their presence. Catbirds are the jazz artists of the bird world, their improvised song a mishmash of short notes and phrases interspersed with pauses, with the odd cat-like call randomly thrown in.

BIRDS | OF THE SHORELINE

Cedar Waxwing

1 Cedar Waxwings acquire the name-giving, red, wax-like appendages on their wings in their second year, and these increase in number as they age.

2 Waxwings nest in late summer to exploit wild fruit when it ripens, but earlier, they often forage over water, snatching Mayflies as they emerge in their adult flying stage.

OF THE SHORELINE | **BIRDS**

Song Sparrow

The melodic songs of Song Sparrows are often heard along the edges of waterways. In general, sparrows are skulkers, but if you get a glimpse of one, the heavy breast streaks with a central spot are good marks for this species.

White-throated Sparrow

1 A new version of the famous "sweet-Canada-Canada-Canada" song of the White-throated Sparrow has apparently swept across the country; now there is one less syllable in "Canada!"

2 White-throated Sparrows have two morphs — one with black and white head stripes, and the other with brown and tan stripes. A mixed pair has more reproductive success than a pair of a single morph because, regardless of sex, "white stripes" are better at defending nests and "tan stripes" are better at feeding young.

American Redstart

With a black head and bright orange flashes in the wings, sides, and tail, a male American Redstart is a handsome bird. But it takes two years to acquire these colors; one-year-old males look like females, with gray and yellow coloring instead of black and orange. On occasion, these active warblers fly out from alder and willow thickets to pluck insects from the air.

BIRDS | OF THE SHORELINE

Yellow Warbler

Well-named for their overall color, Yellow Warblers are the commonest warblers inhabiting shrubby wetlands. Males have reddish breast streaks, while females sport pale versions of these.

OF THE SHORELINE | **BIRDS**

Chestnut-sided Warbler

Chestnut-sided Warblers forage and nest in shrub-dominated habitats. The males sing two songs. The "*please, please, pleased to meetcha*" song with an emphatic last syllable is an "accented" song that serves to attract females. An "unaccented-ending" song announces the male's territory and is used in aggressive encounters with other males.

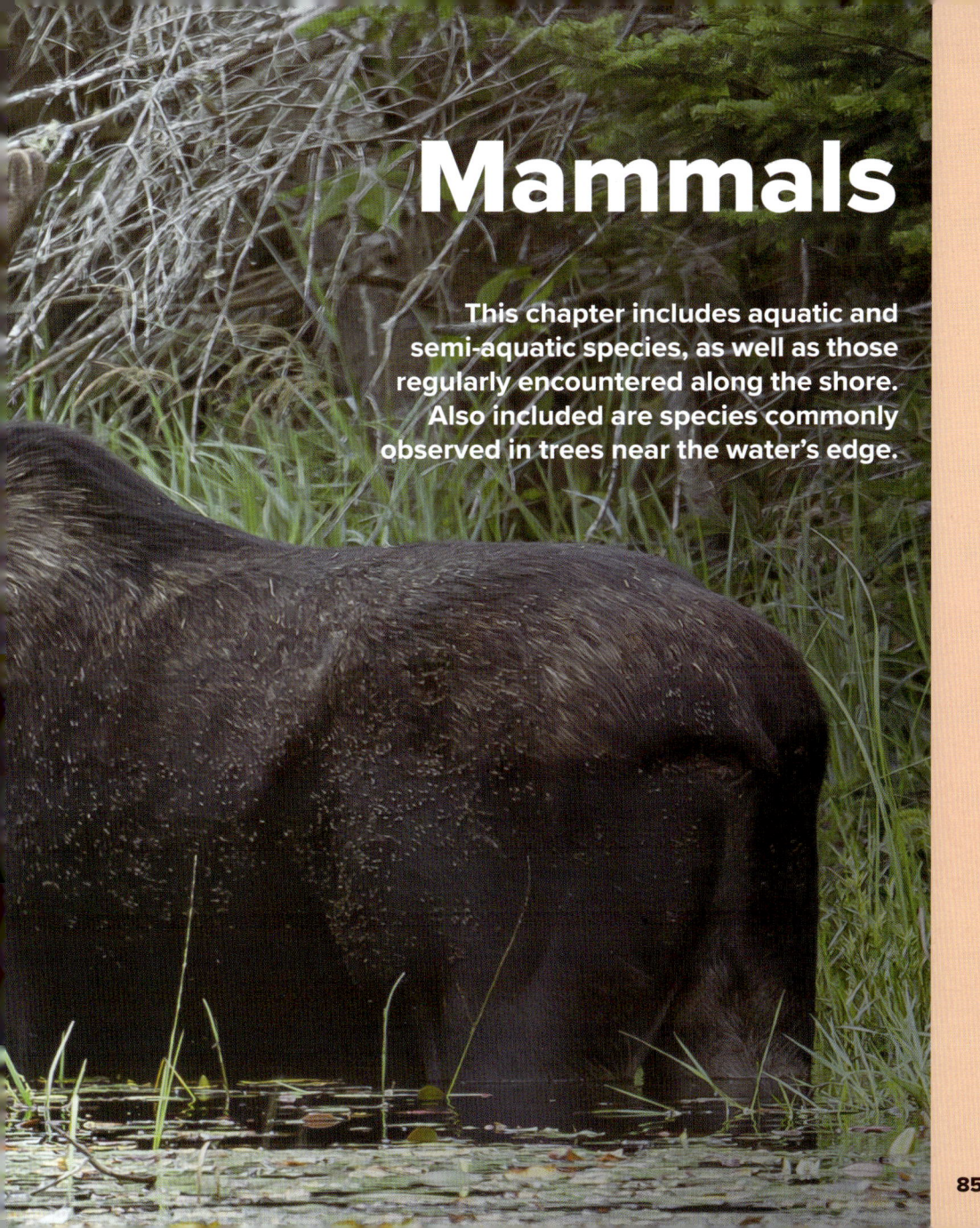

Mammals

This chapter includes aquatic and semi-aquatic species, as well as those regularly encountered along the shore. Also included are species commonly observed in trees near the water's edge.

MAMMALS

River Otter

1 When River Otters are out of the water, one can see their short legs and webbed feet that make them masters of the water. When on the move, these aquatic weasels ("freshwater seals") will suddenly dive and vanish for several (reportedly up to eight) minutes.

2 Otters are curious animals. If you simulate their snort by blowing through your lips to make them flap, they often come in for a closer look, raising their heads high from the water.

American Mink

American Mink are semi-aquatic weasels that patrol the shoreline for frogs and other prey and dive to capture fish. They will take prey larger than themselves, with Muskrats being a notable example.

MAMMALS

Raccoon

Raccoons forage along the water's edge from dusk to dawn for frogs, crayfish, and other edible items. Their front paws contain huge numbers of tactile sensory cells, which is why they grope their food so much — a behavior that gave rise to the myth that they wash their food before eating it. A Raccoon simply "sees" its food better with its remarkable sense of touch.

MAMMALS

Black Bear

1 Black Bears are adept at crossing waterways, either by swimming or using beaver dams as bridges across smaller ones.

2 Black Bears usually flee when they see a human, occasionally standing up first to gain a better view of the intruder. Females have longer ears and paler snouts than males.

MAMMALS

Eastern Wolf

1 The Eastern Wolf formerly ranged across eastern North America, from Florida to southern Ontario. It now exists only in the most northerly part of its former range, with the endangered Red Wolf representing the extreme southern vestiges of the former population.

2 Wolves are more commonly heard than seen especially when the pups are at rendezvous sites in late summer. Rendezvous sites are places where the adults leave the pups while they go hunting, returning regularly to feed them.

Eastern Coyote

When wolves were exterminated in most of eastern North America and woodland habitat was replaced by agricultural land, Coyotes moved in from the prairies. There was some crossing with wolves, and today's Eastern Coyotes (not "coywolves!") often look quite wolf-like. Size is not a dependable criterion, but a coyote's snout is narrower than a wolf's and their habitats are different, with wolves requiring large continuous tracts of forest, avoiding the farmland and urban centers where coyotes thrive.

MAMMALS

White-tailed Deer

White-tailed Deer come to the water's edge to drink and eat sodium-rich aquatic plants. Eastern White Cedar, which commonly grows along river and lake shorelines, is one of their main winter foods.

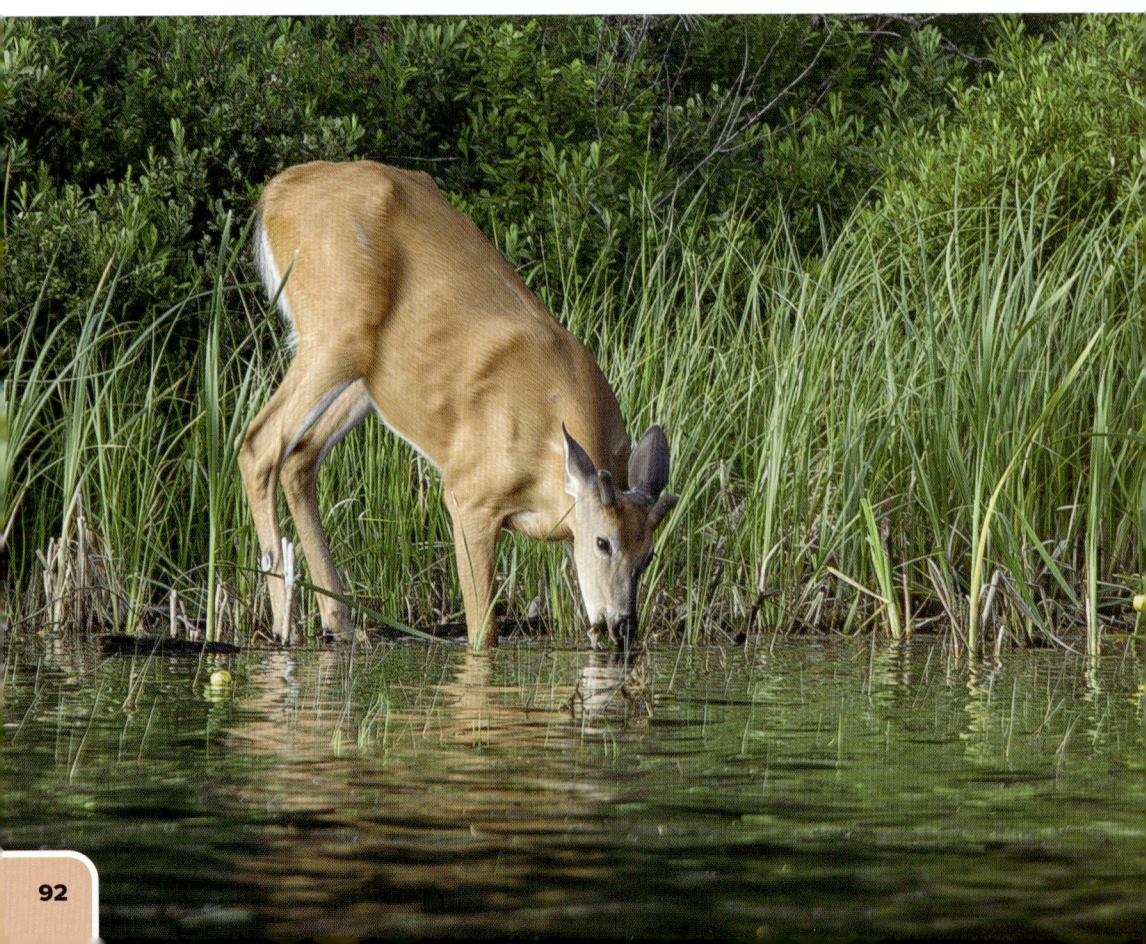

MAMMALS

Moose

Few moments are as exciting as paddling around a river bend and suddenly encountering a huge Moose. Moose do not seek the water to cool down or escape biting flies; they go there to acquire sodium from aquatic plants, especially Watershield, which contains 400 times more sodium than terrestrial plants. Sodium is stored in the stomach's rumen from where it is gradually dispersed throughout the Moose's body over the course of the year.

MAMMALS

Eastern Chipmunk

Eastern Chipmunks are excellent climbers and harvest tree fruit and seeds for consumption during their intermittent winter dormancy, which is spent underground. They also surprise many a paddler, for they swim rather well.

MAMMALS

Red Squirrel

Red Squirrels are usually found in habitats where conifers prevail for the seeds of those trees are an important winter food for them.

MAMMALS

Gray Squirrel

Gray Squirrels live in deciduous forests where tree flowers are devoured in spring and acorns and hickory nuts are stored for the winter. Gray Squirrels come in black and gray color morphs, black being more common in the northern part of their range.

Porcupine

1 Although Porcupines are usually seen in trees, occasionally they can be encountered at the water's edge eating sodium-rich plants. Porcupines cannot throw their quills; an attacker must make physical contact with them before they readily dislodge.

2 Oddly, quills have a greasy coating that contains antibiotics, which might prevent infection if a Porcupine impales itself with its own quills. An alternate theory is that the antibiotics allow a predator to live with a painful but not fatal experience, one that might be imparted to its offspring. Females have longer ears and paler snouts than males.

MAMMALS

Beaver

1 Beavers are commonly seen in early morning or late evening, either swimming across a lake or feasting on water-lilies, one of their favorite foods.

2 When swimming, a Beaver's bulkier head — often held above the water and sometimes all that is visible — wider tail, and larger wake help the observer tell it apart from a Muskrat.

Muskrat

1 Muskrats eat aquatic plants as well as vegetation growing on the shore. They also have a penchant for clams, and large piles of opened shells on the bottom of shallow water are a sign that Muskrats are present.

2 When a Muskrat swims, its vertically flattened tail moves like a snake, helping propel it through the water. The tail's shape also helps distinguish it from a Beaver, as does its smaller size. Additionally, when a Muskrat is swimming, its body is highly visible while a Beaver's head might be all that is seen above the water.

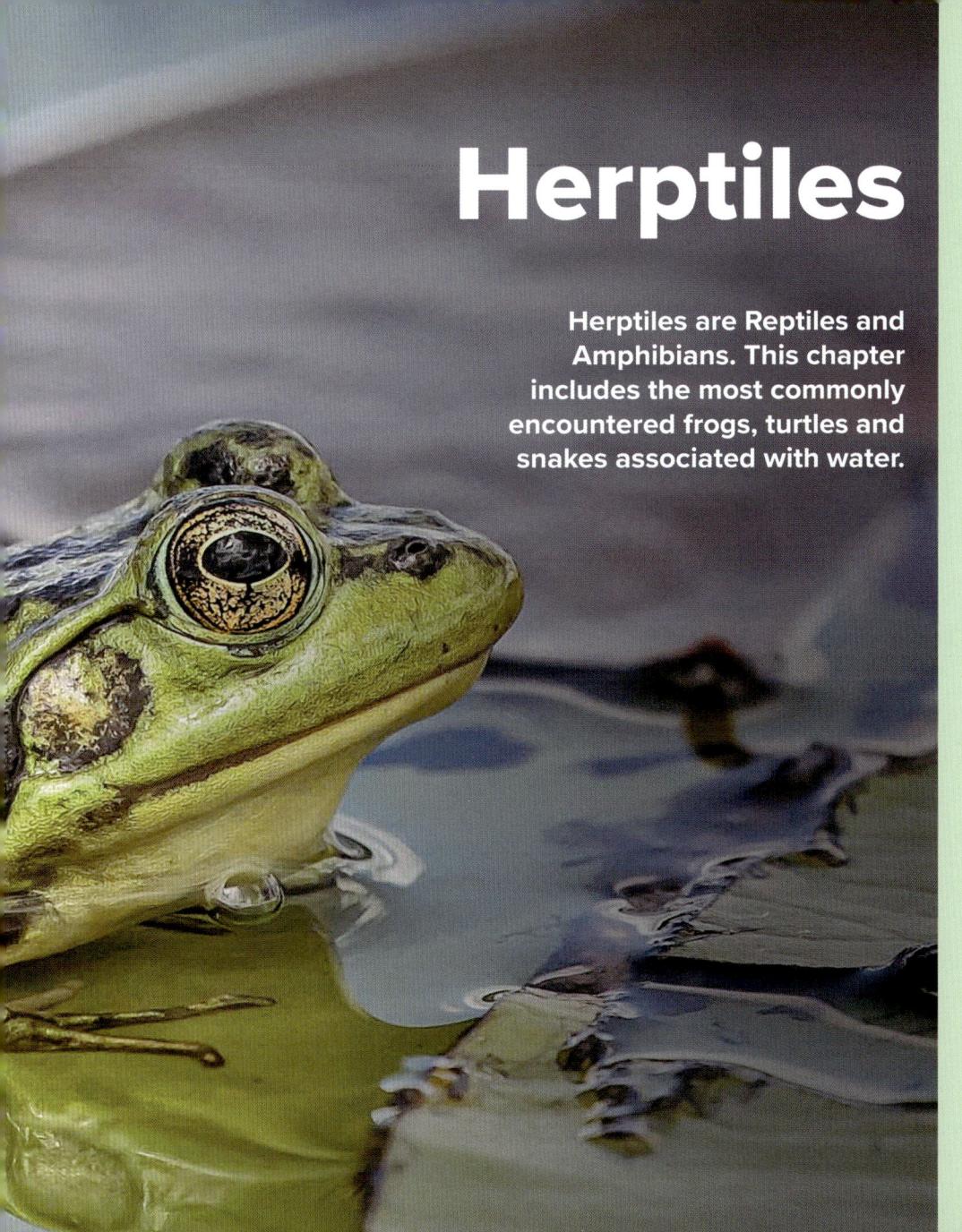

Herptiles

Herptiles are Reptiles and Amphibians. This chapter includes the most commonly encountered frogs, turtles and snakes associated with water.

Bullfrog

Sadly, the baritone songs of male Bullfrogs are currently being heard less commonly on warm summer days. The largest frog in the region, a Bullfrog's sex can be told by the size of its tympanum ("eardrum"): the tympanum of a male is larger than its eyes, while the tympanum of a female is the same size as her eye or smaller.

Green Frog

With their bright yellow throats, Green Frogs may look like small Bullfrogs, but the ridge running down both sides of their back readily distinguishes the former from the latter. Their voices are also very different, with a Green Frog sounding like a banjo being plucked.

HERPTILES | FROGS

Mink Frog

1 When alarmed, a Mink Frog releases a name-giving, mink-like musky odor. These heavily mottled frogs inhabit colder waters typical of more northerly regions. A Mink Frog's "song" is like a loud tap; a chorus of males is said to sound like the popping of popcorn.

2 Frogs amplify their songs by forcing their sound into an inflatable throat sac. Mink Frogs have one sac, but some frogs including the Northern Leopard, have two.

FROGS | **HERPTILES**

Northern Leopard Frog

While Northern Leopard Frogs are often found far from water, they are also regularly encountered along shores. Their song sounds like a loud snore. The distinctive round spots visually break up the body into separate parts, a "disruptive pattern" that makes a Leopard Frog hard to see until it moves. When the legs are folded in the sitting position, the blotches line up, enhancing this form of camouflage.

HERPTILES | FROGS

Pickerel Frog

A Pickerel Frog looks not unlike a Northern Leopard Frog, which can be brown as well as green. However, Pickerel Frogs sport square, not round, markings.

American Toad

About the only time you will encounter an American Toad in the water is in spring, when they are mating. At that time, large numbers of toads gather in shallow water, and the piercing trills of males fill the air. In the autumn, toads dig down below the frost line, where they remain dormant in the ground until the warmth of spring awakens them.

Northern Watersnake

1 Most snakes encountered in Northeastern waters swimming or sunning along the shore are likely to be a Northern Watersnake. As the name suggests, these snakes are excellent swimmers and divers that eat small fish and frogs.

2 Female watersnakes are larger than males, and young snakes are more brightly patterned than older individuals, which can appear all black.

3 When a snake's eyes go cloudy, it is a sign that it is about to shed its skin.

SNAKES | HERPTILES

2

3

109

Eastern Gatersnake

The Eastern Gartersnake is the most common snake in the Northeast and often hunts frogs by the water's edge. This species is highly variable in appearance, and while most individuals have green and white stripes running down the length of their body, some are boldly checkered and individuals in the west have red sides.

Eastern Ribbonsnake

Eastern Ribbonsnakes are semi-aquatic; hence, they are usually encountered not far from water. They superficially resemble a slim, vividly marked gartersnake, but the white chin and white spot in front of the eye are diagnostic marks.

Painted Turtle

Painted Turtles are commonly seen basking on rocks or logs to raise their internal body temperature and drive leeches from their skin. Like other turtles, they lay their eggs in the ground in late spring, with summer temperatures dictating the sex that will emerge in the fall. However, unlike other turtles, some hatchlings remain in the soil overwinter, with half their body water turning to ice, a feature known as freeze tolerance. This amazing feat is accomplished only by hatchlings; all other ages spend the winter at the bottom of shallow waters where temperatures remain above the freezing mark.

TURTLES | **HERPTILES**

Snapping Turtle

1 Snapping Turtles can live 100 years or more. Although they don't bask throughout the summer as Painted Turtles do, prior to laying, to enhance egg development, female snappers bask to raise their internal body temperature.

2 When basking near the water's surface, Snapping Turtles stretch out their legs to expose more of their body surface area to the sun. This position is known to allow Painted Turtles to glean leeches from their skin!

HERPTILES | TURTLES

Map Turtle

1 Map Turtles — named after the topographic map-like lines on their shells — generally inhabit larger bodies of water. They might be overlooked as Painted Turtles, but they lack red on their sides and have a keel atop their carapace (upper shell).

2 Map Turtles often bask in large groups on rocks and logs.

Blanding's Turtle

1 With their bright yellow throats and domed shells Blanding's Turtles are distinctive. These generally shy turtles prefer smaller bodies of water but often make long journeys overland to reach egg-laying or wintering sites.

2 Unlike most turtles, Blanding's Turtles can partially close the front of their shells after withdrawing their head and legs inside for protection.

Insects

Insects are the most common animals encountered on canoe or kayak outings. Main categories are Damselflies, Dragonflies and Flies. Other Insects include species seen on the water or flying above it, such as Caddisflies, Mayflies and Beetles.

INSECTS | DAMSELFLIES

River Jewelwing

Damselflies and dragonflies are in the order Odonata, which means "toothed ones," a reference to their serrated mandibles. All of them are predators; damselflies are typically smaller, with thinner bodies and their eyes well separated. Damselflies typically close their wings above their backs when perched. Two-toned wings readily separate the River Jewelwing from Ebony, which sports solid black wings. Although the habitat of the two species overlaps, River Jewelwings often the frequent edges of waterways with a little more flow.

DAMSELFLIES | **INSECTS**

Ebony Jewelwing

1 Jewelwings are giant damselflies, with Ebony Jewelwing arguably being the most beautiful. They frequent smaller waterways, slowly flitting over the water before returning to their shoreline perches. Males perform slow courtship flights over females, who reveal their decision by holding their wings flat to say: "No" or clapping them over their backs to signal: "Come see me!"

2 Female Ebony Jewelwings sport white in the outer corner of their wings. They lay eggs in aquatic vegetation, placing them in slits that they cut with a knife-like appendage on the tip of their abdomen.

INSECTS | DAMSELFLIES

Violet Dancer

Violet is an unusual color in nature, and male Violet Dancers are the only violet damselflies you will encounter. They usually perch on rocks or shores but will occasionally land on a canoe or kayak.

DAMSELFLIES | **INSECTS**

Powdered Dancer

As they age, Powdered Dancers develop a white dusting known as pruinosity. These common damselflies often perch on rocks and appear all white when they skim over the water's surface.

INSECTS | DAMSELFLIES

Aurora Damsel

These Aurora Damsels have coupled but not yet joined in the distinctive "heart" configuration in which mating occurs. Unlike most damselflies, apart from the Spreadwings, the wings of an Aurora Damsel remain partly open when perched. Also, the yellow sides and sold black back separate this species from the similar-looking Bluets.

DAMSELFLIES | INSECTS

Skimming Bluet

Bluets are a large group of damselflies in which males are typically blue and black, with differing proportions of each color. The Skimming Bluet, which shows more black than blue (a "black-type" bluet) is often seen perched on floating vegetation.

INSECTS | DAMSELFLIES

Tule Bluet and Boreal Bluet

1 The Tule Bluet is an "intermediate type," showing equal amounts of blue and black. These damselflies are often encountered flying low over open water.

2 The Boreal Bluet, one of the most common Bluets, is a "blue type," as that color dominates.

DAMSELFLIES | **INSECTS**

Stream Bluet

The Stream Bluet is the most commonly encountered Bluet along faster-moving water.

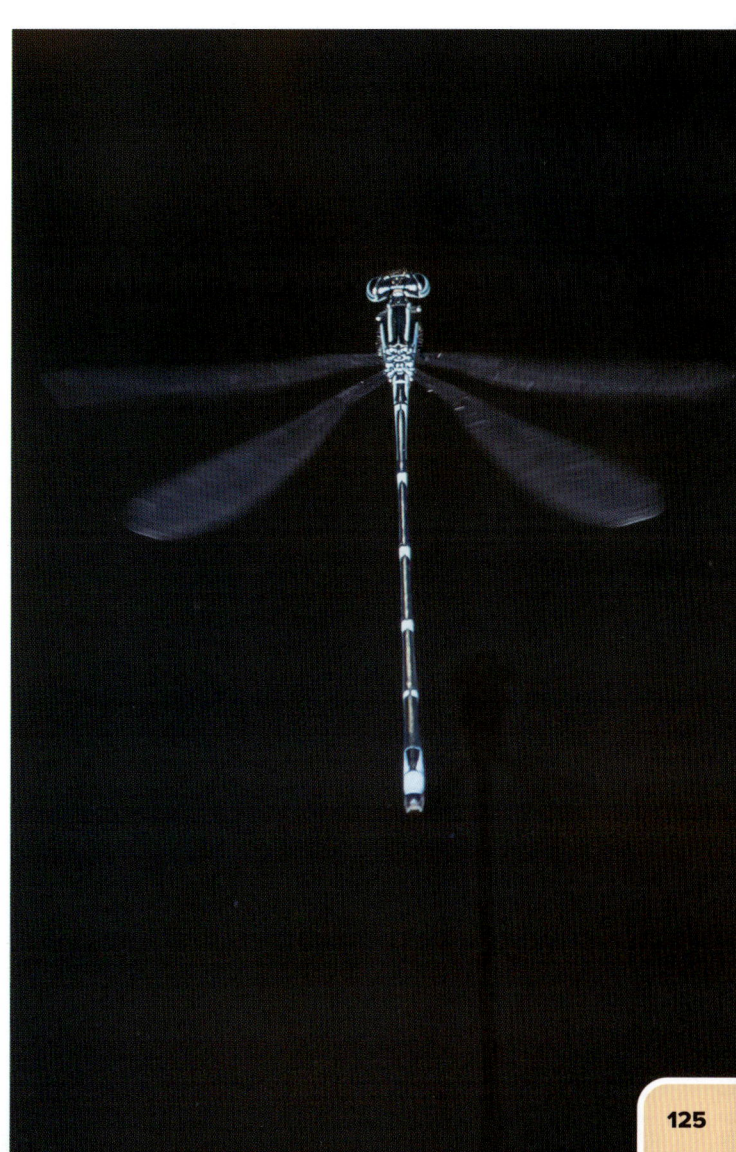

INSECTS | DAMSELFLIES

Orange Bluet and Vesper Bluet

1 Not all bluets are blue! Orange Bluets are most active later in the day, flying over deeper water. This pair is mating in the unique Odonate position known as the "wheel," which might better be called the "heart."

2 A favorite of Daniel Craig, the Vesper Bluet usually does not make an appearance until after the sun has set, landing on floating vegetation or, here, a shed pine flower.

DAMSELFLIES | **INSECTS**

Northern Spreadwing

As its name suggests, this Northern Spreadwing and others in its group characteristically hold their wings half-open when perched, not closed over the back like other damselflies, and never flat like dragonflies. Spreadwings can be difficult to identify as a species; males often require the scrutinization of a male's claspers (species-specific appendages that project from the abdomen tip and get inserted behind a female's head during mating).

INSECTS | DRAGONFLIES

Common Green Darner

1 The Common Green Darner (here, a male) is migratory, arriving in spring to the Northeast from the far south.

2 Upon arrival, these huge dragonflies mate and lay eggs, as this female is doing. The ensuing generation arises from rapidly developing nymphs and migrates south. It is their offspring that return the following year.

Lake Darner

Sometimes one must catch a dragonfly to see its distinctive features, including the shape of its thoracic markings. The proper way to safely hold a dragonfly like this Lake Darner without hurting it is to pinch its wings over its back.

INSECTS | DRAGONFLIES

Canada Darner

Darners are large dragonflies that patrol lake edges but also gather inland in large feeding swarms late in the day. The shape of the large thoracic markings are species-specific, identifying this dragonfly as a Canada Darner.

Shadow Darner

Shadow Darners often hunt along the shore, flying low and late into the evening when shadows grow long.

INSECTS | DRAGONFLIES

Fawn Darner

Fawn Darners fly low along the shoreline of streams, rivers, and lakes, methodically investigating rocks, logs, and branches for insect prey, or laying eggs — as this female is doing. The Ocellated Darner looks similar but has larger yellow markings on the abdomen and prefers rocky waters with a current.

Black-shouldered Spinyleg

1 After spending at least a year in its aquatic stage, a dragonfly nymph crawls out of the water. As its exoskeleton splits open, it emerges as a flying dragonfly. A newly emerged Black-shouldered Spinyleg is yellow, but as it ages, it changes to olive green with its eyes turning blue. A male Spinyleg, like most other Clubtails, sports a larger abdominal "club" than the female, and also has two sets of claspers that hold the female by the back of her head during mating.

2 In flight, a Black-shouldered Spinyleg characteristically points its abdomen upward (unlike the Prince Baskettail flying beneath it).

Dragonhunter

Named for its habit of eating other dragonflies, the Dragonhunter has a huge body but small head. In flight, the males curl their clubbed abdomen downward, giving them a distinctive sideways "J" profile.

DRAGONFLIES | **INSECTS**

Lilypad Clubtail

Most male Clubtails have the final segments of their abdomen enlarged into a "club," and their eyes, unlike those of other dragonflies, do not touch each other. Lilypad Clubtails are well-named for they use the floating leaves of Water-lilies as observation posts from which they launch attacks on flying insects.

INSECTS | DRAGONFLIES

Boreal Snaketail

Boreal Snaketails are common in northern regions, and regularly encountered along fast-flowing streams and rivers as far north as Hudson Bay.

DRAGONFLIES | **INSECTS**

Rusty Snaketail

A beautiful Clubtail, the Rusty Snaketail is regularly encountered perched on rocks in shallow rapids.

INSECTS | DRAGONFLIES

Lancet Clubtail

Lancet Clubtails are a small dragonfly. While some male clubtails are difficult to identify without looking at the claspers at the end of their clubs, Lancets can be distinguished by having yellow markings on top of all of their abdominal segments. This male is enjoying a moth for dinner.

Stream Cruiser and Swift River Cruiser

1 Stream Cruisers, in addition to streams also "cruise" the edges of rivers and lakes searching for prey or defending territories.

2 Swift River Cruisers are huge, fast flying dragonflies that "cruise" the shores of rivers and lakes. When not flying, they hang diagonally from branches.

INSECTS | DRAGONFLIES

Common Baskettail

Baskettails have a unique way of laying eggs. The female extrudes her eggs, while perched, into a pouch-like structure at the end of her abdomen. She then flies low over the water, dragging her "basket" and washing out the eggs string-like. Baskettails look very much alike, but Common Baskettails usually have a large dark patch at the base of the hind wings.

DRAGONFLIES | INSECTS

Prince Baskettail

Prince Baskettail is the largest Baskettail and often flies high over lakes. Generally a slow flier, its three dark markings on each wing allow for easy identification.

INSECTS | DRAGONFLIES

Eastern Amberwing

Less than an inch in length, an Eastern Amberwing is one of the smallest dragonflies in the world. However, with its blazing orange-gold wings, a male is certainly one of the most beautiful ones! These tiny predators frequent slow-moving rivers, where they perch on logs, rocks, and floating vegetation, waiting for a meal to fly by. Their bright color and erratic, rapid flight have been suggested to mimic that of wasps, a feature that might make them less desirable as a meal for birds.

DRAGONFLIES | INSECTS

Eastern Pondhawk

Female Eastern Pondhawks have a bright green thorax, but the thorax and abdomen of a male is blue. Like other dragonflies, Pondhawks are ferocious predators, with this female enjoying a Hairstreak, a tiny butterfly.

INSECTS | DRAGONFLIES

Blue Dasher

On hot days dragonflies like this Blue Dasher will raise their abdomen high in the air (a position called obelisk) to reduce the surface area of their body being heated by the sun. The raised abdomen can also act like an umbrella, casting shade on the rest of the dragonfly, keeping it cooler.

DRAGONFLIES | INSECTS

Dot-tailed Whiteface

The Whitefaces are small, dark dragonflies with, of course, a white face! The Dot-tailed Whiteface is a great name only for the males of this species, which often perch on leaves of aquatic plants; the females closely resemble other members of their group.

INSECTS | DRAGONFLIES

Frosted Whiteface

The Frosted Whiteface is a common sight perched on boggy edges of small lakes and ponds. In addition to the white face, the males develop a whitish bloom (pruinosity) on the base of their abdomen. The much larger Chalk-fronted Corporal shows pruinosity on its abdomen and also above its wings, and has a dark face.

DRAGONFLIES | INSECTS

Cherry-faced Meadowhawk and White-faced Meadowhawk

1 Meadowhawks are small dragonflies that fly in late summer. The males of most species have red bodies. The Cherry-faced Meadowhawk looks very much like the Ruby Meadowhawk but that species' face varies from orange to brown, not cherry-red.

2 The White-faced Meadowhawk is the only Meadowhawk with a white face.

147

Autumn Meadowhawk

One of the last dragonflies to fly each year, the Autumn Meadowhawk has distinctive straw-colored legs.

Twelve-spotted Skimmer

One of the larger dragonflies, the wing pattern of Twelve-spotted Skimmer males is distinctive, but the females look very much like female Common Whitetails. Size is a feature (Twelve-spotted females are larger), but a clincher is the series of horizontal marks on the sides of the abdomen: these are slanted in Whitetails.

INSECTS | DRAGONFLIES

Common Whitetail

1 Male Common Whitetails develop a white coating known as pruinosity on the abdomen, which — along with their half-dark wings — makes them easy to recognize.

2 Female Common Whitetails have three dark spots on their wings like Twelve-spotted Skimmers, but, unlike their larger relative, the pale markings on the sides of their abdomen are slanted and do not form a straight line.

Widow Skimmer

Skimmers are heavy-bodied dragonflies that often spend their larval stage in still or slow-moving water. One suggested origin of the Widow Skimmer's name is that the dark base of the wings resembles a widow's black shawl.

INSECTS | DRAGONFLIES

Slaty Skimmer

1 Older male Slaty Skimmers are distinctive for their color.

2 Young male and female Slaty Skimmers look like a very different species.

Chalk-fronted Corporal

A common dragonfly of shallow lake edges on the Canadian Shield, Chalk-fronted Corporals sometimes cover shoreline rocks when they bask to raise their body temperature. They are often encountered perched on the ground on trails and portages.

INSECTS | DRAGONFLIES

Four-spotted Skimmer

While many dragonflies have quite different looking sexes (sexual dimorphism), apart from their genitalia, male and female Four-spotted Skimmers look pretty much the same. In Europe this species is known as the Four-spotted Chaser.

Giant Eastern Crane Fly

Crane Flies look like giant mosquitoes but have no fear — they don't bite, and in fact, some species don't even eat. The Giant Eastern Crane Fly is one of the biggest and prettiest species; its predaceous larvae live in the moss-covered edges of streams or in their mucky bottom. As their incredibly long legs break off easily, it is not rare to see one with fewer than six legs.

INSECTS | FLIES

Phantom Crane Fly

Often the only glimpse of a Phantom Crane Fly you get are its white leg segments as it slowly drifts between plants along the water's edge. Although these flies have wings, they float using air-filled sacs on their outstretched legs to catch the slightest of breezes, just like balloons.

FLIES | **INSECTS**

Mosquito

Unlike other biting flies, female Mosquitoes stick a needle-like proboscis into your skin. The reason a Mosquito "bite" makes you itch is because, like that of other flies, their saliva contains an anticoagulant to keep the blood flowing. It is a reaction to this that causes the itch.

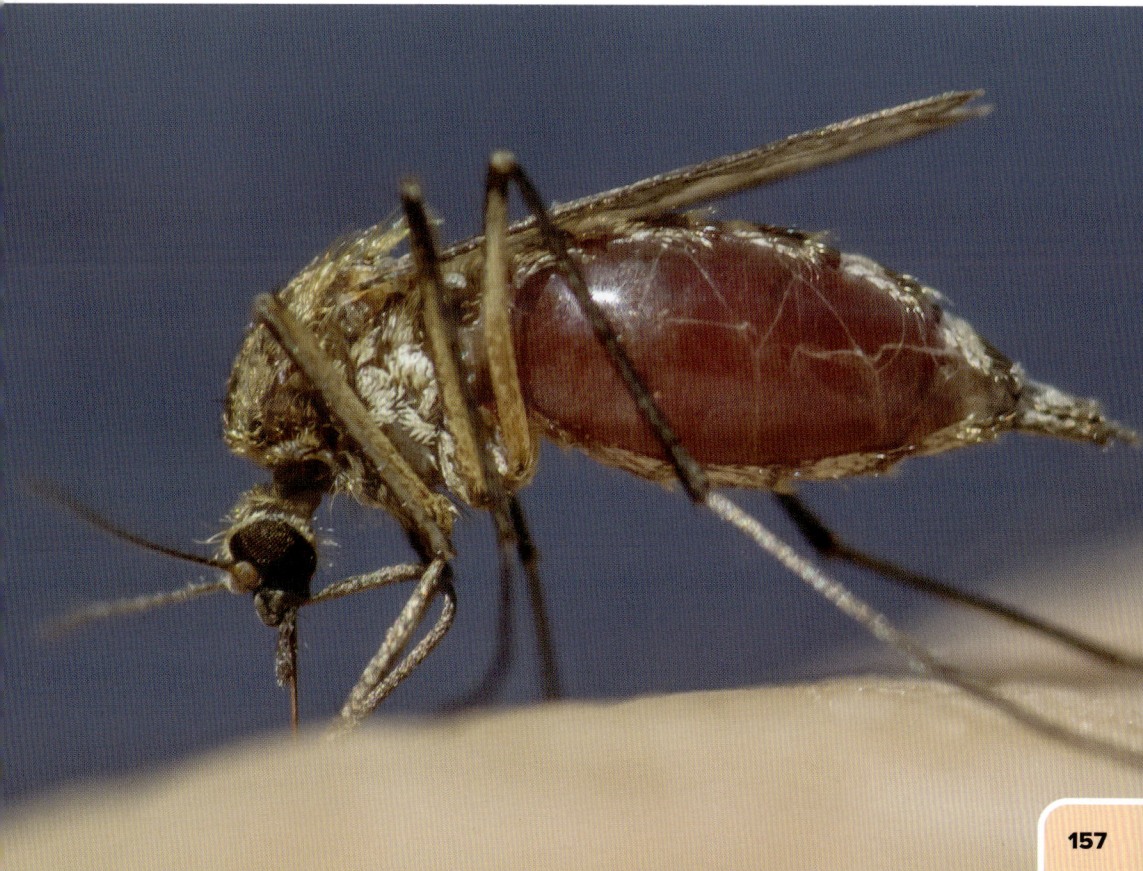

Long-legged Fly

Long-legged Flies are small predatory flies that often perch on floating leaves of aquatic plants or the shore. They feed by masticating their prey — often a Midge, as seen here — and then sucking up the liquid, discarding the empty carcass.

FLIES | **INSECTS**

Black Fly

Anyone who has paddled in northern regions is more than familiar with female Black Flies. Using sensors on their antennae and legs to detect carbon dioxide and heat from as far away as 35 meters, these minute flies quickly swarm a target. Their razor mandibles snip through skin and anticoagulants keep blood flowing, providing the females with extra protein for egg development.

INSECTS | FLIES

Deer Fly

1 Deer Flies replace Black Flies as friendly visitors as summer passes. There are numerous species, some identifiable by their colorful eye patterns. The eyes reveal the sex of Deer and Horse flies: those of males touch each other, while those of females do not.

2 Deer Flies often visit flowers for nectar, a popular food. As in most other biting flies, the females extract blood only for egg development.

Horse Fly

1 Female Horse Flies, like female Deer Flies, employ two sets of mouthparts to access your blood. One pair (the maxillae) has serrations and act like saws, moving in and out, while the other (the mandibles) cut like scissors.

2 Male Horse Flies hover in sunlit forest openings, their abdomens glowing like lanterns, waiting to pursue passing females or chase away rivals with speeds reaching 90 kilometers per hour!

Moose Fly

You never know how many flies are on a Moose until it enters the water. Then, suddenly a living cloud appears, only to quickly settle back on the Moose's rump. Don't despair — Moose Flies don't bite humans. Unlike most biting insects, both male and female Moose Flies dine on their host. Their entire life cycle is tied to Moose; females lay their eggs in Moose droppings, so guess what the larvae eat!

FLIES | **INSECTS**

Stable Fly

1 Unlike Tabanids (Deer and Horse flies), both sexes of Stable Flies draw blood. They are particularly skilled at biting the ankles and feet of paddlers. A Stable Fly's bayonet-like proboscis is tipped with rasping teeth that cut the skin.

2 Few paddlers will admire Stable Flies for their looks, but they actually are quite attractive — much prettier than their close relatives, the House Fly.

INSECTS | OTHER INSECTS

Caddisfly swarm

Caddisflies are four-winged insects that spend their larval stage in the water, many living in cases they build from nearby materials such as plant debris or sand. When they emerge from the water as adults, the males form mating swarms such as this one.

OTHER INSECTS | **INSECTS**

Zebra Caddisfly

There are different families of Caddisflies, many of which contain small species. The Zebra Caddisfly belongs to the Hydropsychidae, the Net Spinning Caddisflies, a name that arises from the elegant silk nets build by the larvae to filter food from the water.

INSECTS | OTHER INSECTS

Long-horned Caddisfly

If you see a swarm of minute dark insects whirling around at dizzying speeds just above the water's surface, you might be seeing a mating group of Mayflies. When these Long-horned Caddisflies (about 5 millimeters in size) were captured on camera, they looked like fairies.

OTHER INSECTS | **INSECTS**

Little Stout Crawler Mayfly

One of the best ways to see what Mayflies are emerging from a waterway is to look at spider webs on bridges or the shore. At about four millimeters, these Little Stout Crawler Mayflies are one of the smallest Mayflies, but their numbers make them an important food not only for spiders, but also for birds like Cedar Waxwings.

INSECTS | OTHER INSECTS

Brown Drake

Mayflies live as aquatic larvae for a year before they float to the surface and emerge from the shed skins as flying adults. They typically emerge in huge numbers, their bodies decorating docks and other buildings near water. These Brown Drakes, like other Mayflies, are commonly called Shadflies.

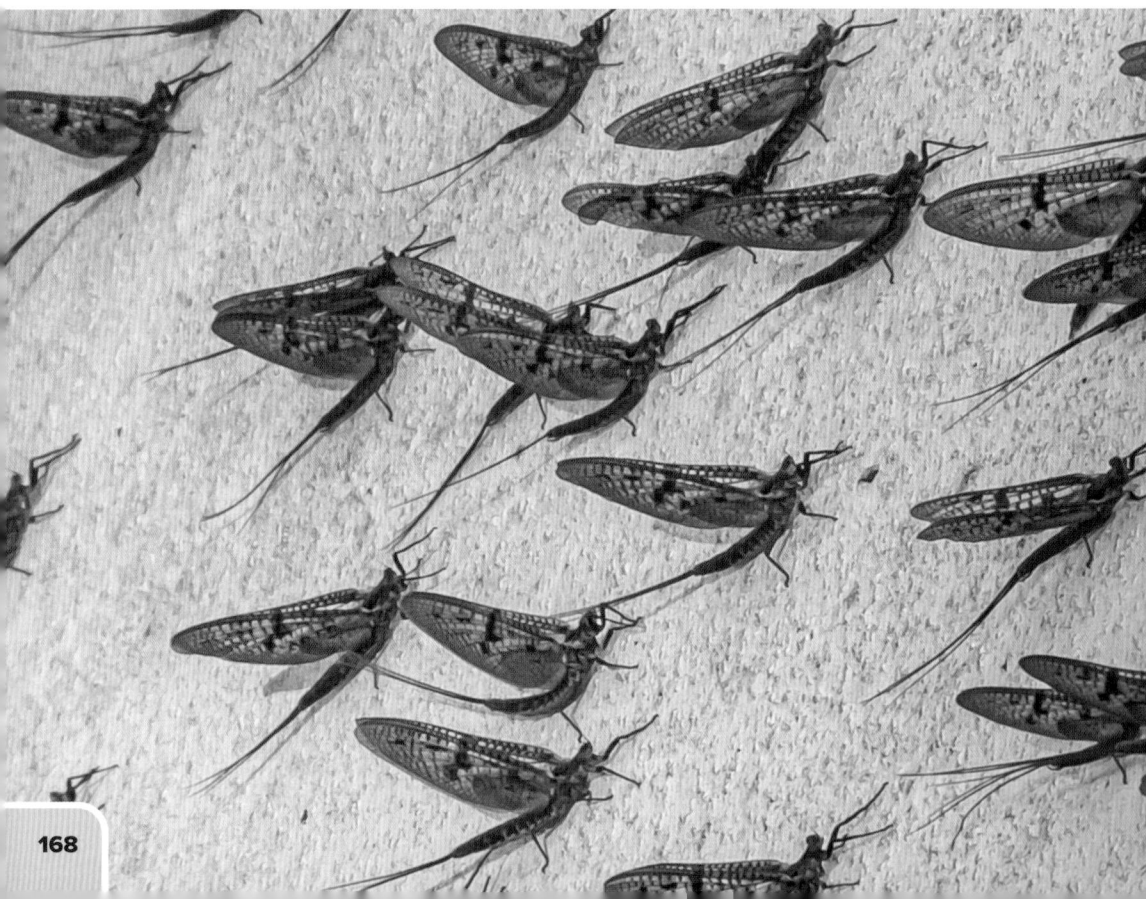

OTHER INSECTS | **INSECTS**

Giant Mayfly

Mayflies lay their eggs on or in the water, their spent bodies adorning it after their life objective has been fulfilled. Lacking mouthparts, adults like this Giant Mayfly have a very short life expectancy: females often only a few hours and males one to two days.

INSECTS | OTHER INSECTS

Stonefly

1 Adult Stoneflies hold their wings flat over their backs. Like Mayflies and many other insects with an aquatic larval stage, adult Stoneflies do not eat.

2 Stoneflies spend most of their life in the water as gill-breathing nymphs. Their shed exuviae adorn rocks onto which they crawled before emerging as flying adults.

Whirligig Beetle

Whirligigs are predatory aquatic beetles that are encountered in large groups, which whirl about like miniature bumper cars without ever hitting each other. The upper part of their body stays above the water due to a fringe of water-repellant (hydrophobic) hairs. Whirligigs possess two sets of eyes, one that surveys the world above the water and a submerged pair for viewing beneath the surface.

INSECTS | OTHER INSECTS

Aquatic Leaf Beetle

Aquatic Leaf Beetles (Donacia) are herbivores that sit atop the floating leaves of Water-lilies and other aquatic plants they eat, quickly flying to a nearby leaf when alarmed. Their larvae dine on submerged stems and roots, breathing oxygen they steal from their host through a tooth-like "spur" they stick into the plant's air chambers.

OTHER INSECTS | **INSECTS**

Water Strider

Predatory insects, Water Striders swiftly skate across the water's surface, using its surface tension and thousands of hydrophobic (water-repelling) hairs per square millimeter on their legs and bodies to prevent them from becoming water-logged.

INSECTS | OTHER INSECTS

Backswimmer

Never pick up a Backswimmer. These predaceous insects (true bugs) have a needle-like mouthpart for injecting poison into their prey that gives you a jolt that feels like a wasp sting! Like their name suggests, Backswimmers swim belly-up. Water Boatmen, aquatic bugs that eat algae, also have oar-like legs for swimming, but swim back-side-up and lack that injection tool. Both groups of aquatic bugs are dark on one surface and light on the other, for camouflage both from below and from above. Of course, the color of each surface is reversed in the two groups.

Polymorphic Pondweed Moth

There are several species of small white moths whose caterpillars eat the leaves of aquatic plants. This Polymorphic Pondweed Moth is one of the more nicely marked ones found sitting on the floating leaves of water-lilies.

Other Animals

Other Animals include non-insect invertebrates such as Freshwater Sponges, Leeches and Spiders.

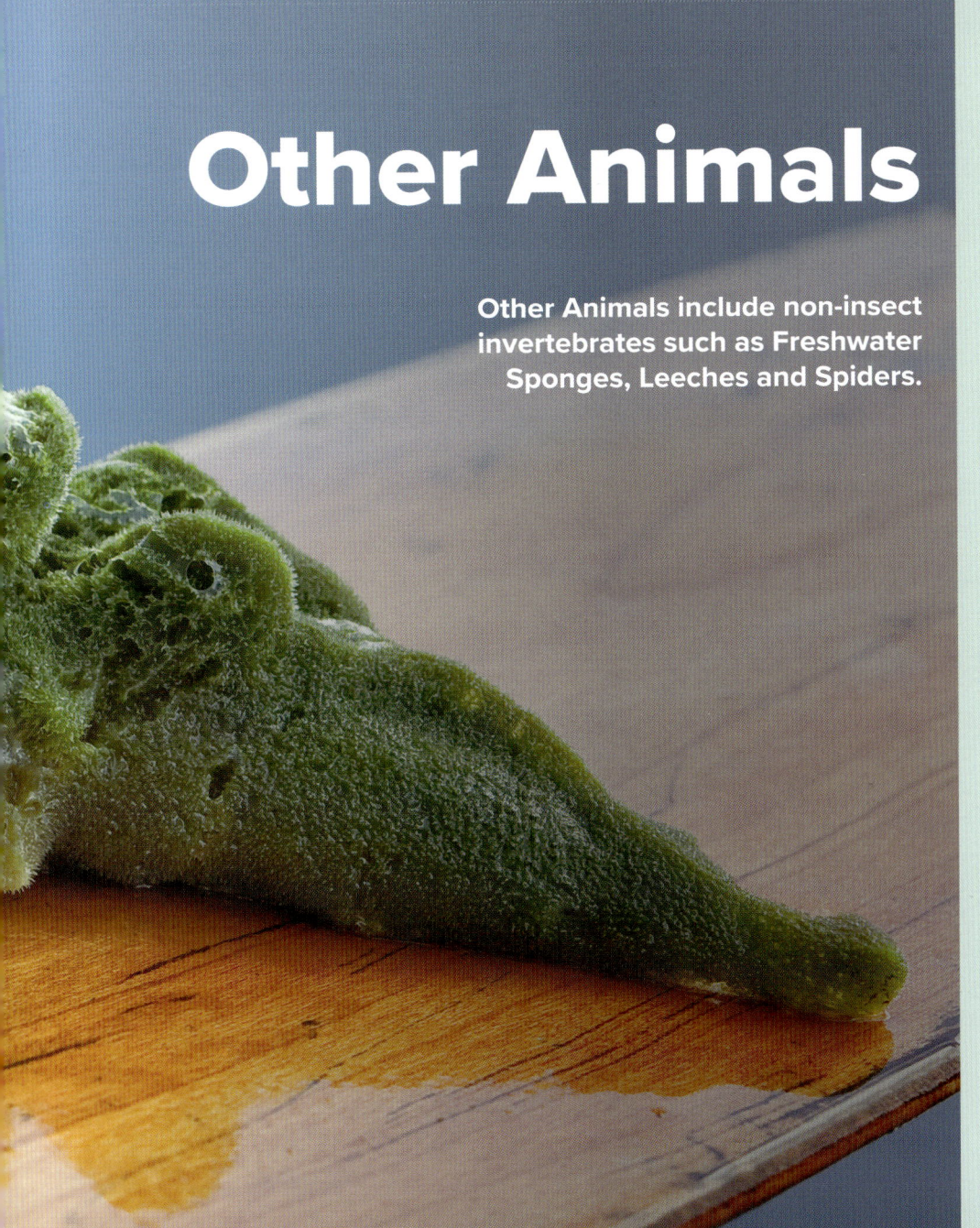

OTHER ANIMALS

Freshwater Sponge

It may come as no small surprise that the green finger-like projections or mounds on rocks or branches beneath your watercraft are not plants or algae, but animals! Freshwater Sponges are colonial animals that glean food from water as it passes through them. Because sponges filter out whatever the water carries, their presence indicates a healthy waterway.

Rusty Crayfish

There are fewer than a dozen species of Crayfish in northeastern North America. They are omnivorous and, while some dig burrows in moist ground, most live in streams and lakes. This Rusty Crayfish, a non-native species, was caught by a Ring-billed Gull but apparently it wasn't to its liking.

OTHER ANIMALS

Leech

Few are the paddlers who haven't encountered Leeches on a trip. While they do not harm us, people have a strong aversion to these much-maligned animals. To dislodge a Leech, don't yank it off and definitely do not put salt on it! Use something thin such as a credit card to gently pry its mouthparts from your skin. Some females carry their young on their underside until they find a host, then they all dine — something akin to a family picnic!

OTHER ANIMALS

Smooth Turtle Leech

Many leeches are specific to other animals, and a few are predatory. The Smooth Turtle Leech is, as the name suggests, found on turtles, especially Snapping Turtles. Although leeches attach to a turtle where their host can't dislodge them, they are still not safe; Painted Turtles have been observed gleaning them from the underside of Snapping Turtles!

OTHER ANIMALS

Dark Fishing Spider

Fishing Spiders, such as this Dark Fishing Spider, are semi-aquatic and dive to catch prey as large as tadpoles and small fish. These large spiders are often seen running across the water or sitting by the water's edge.

OTHER ANIMALS

American Nursery Web Spider

Like other members of the Nursery Web Spider family (Pisauridae), female American Nurseryweb Spiders build a name-giving web in which they place their egg sac. The female remains to guard the sac and ensuing young. The defence can be aggressive; try lightly jiggling a nursery web with your finger!

Signs of Animals

When paddling, you often pass by signs revealing the presence of animals. Some of these are quite obvious; others, less so. As it would require an entire book to cover every sign you might come across, here is but a small sample of things for which to keep an eye out for during your next outing.

SIGNS OF ANIMALS

River Otter

There may seem to be no sign as to what made this disturbance in the mud, but there is one important clue on the far right. The sliding mark reveals that a River Otter had slid down the slope into the water. The amount of dug-up mud suggests that a family group made a stop on the river bank.

Scent Mound

Piles of fresh mud pushed up on the edges of waterways or on rocks in the water are Beaver scent mounds. Beavers urinate on these, adding anal gland secretions and castoreum to the mound, which serve as personalized "No Trespassing" notices. The moist, raised mud piles help broadcast these chemical advertisements more widely.

SIGNS OF ANIMALS

Flowerless Stalks

If you see a lot of Yellow Pond-lilies that appear to be decapitated, that is because Beavers have selectively eaten their flowers — a favorite food.

Feeding Platform

Flattened vegetation by the water's edge could be caused by River Otters playing or eating (especially if crayfish or fish remains are present), but if debarked sticks are visible then you are seeing a Beaver's feeding platform — a place where it also spends time grooming its fur. If sticks or half-eaten vegetation are not present, then it is likely used only for grooming.

SIGNS OF ANIMALS

Beaver Lodge

Beaver lodges are made out of sticks and mud. Occupied ones have fresh mud and/or freshly cut branches still holding their leaves applied regularly. Trails leading to the lodge through the aquatic vegetation are also signs that someone is home.

Muskrat Lodge

Muskrat lodges are constructed primarily from vegetation, especially cattail leaves and stems, and lack the woody material used by Beavers in their castles. Canada Geese often place their nests atop either type, some still occupied by their builders.

SIGNS OF ANIMALS

Insect Feeding

As summer progresses, many floating leaves acquire trails chewed by feeding insects. Leaf Beetles and their grubs as well as moth larvae chew patterns in the leaves, sometimes creating abstract works of art. Do you see Cinderella in this offering?

Scat

Mammal scats often adorn rocks and logs in waterways. Scat pointed at the ends is typically that of a Mink or River Otter, with those of Mink being much smaller. This Mink scat appears to contain plenty of tubes that might be empty cases of larval Case-building Caddisflies.

SIGNS OF ANIMALS

Clam Sign

Doodles in wet sand are signs that clams were on the move, seeking deeper water when the level of a shallow lake began to drop.

Turtle Nest

Scattered golf ball-sized egg shells near a hole recently dug in the soil reveal that a turtle nest was depredated. The scats reveal that a Red Fox was responsible for this scene.

Dragonfly Nymph Case

The shed skins (exuviae) of dragonflies, as well as those of damselflies and Stoneflies decorate rocks, shoreline vegetation, and even tree trunks along waterways, a testament to how so many flying insects depend on water for their larval development.

SIGNS OF ANIMALS

Trumpet Net Caddisfly

When passing over shallow, moving water, you often see trumpet-shaped nets adorning submerged rocks or branches. These are made by Trumpet Net Caddisfly larvae, which reside in the narrow bend of their silk structure, and eat organic material filtered from the water flowing through their amazing net.

SIGNS OF ANIMALS

Tracks

1 Exposed mud along the edges of waterways is like a hotel register, recording the comings and goings of wild visitors. Each rain turns the page, revealing a new one for the next set of arrivals to sign. This page was autographed by the hand of a Raccoon.

2 Pairs of offset tracks indicate that a weasel scampered across the mud. The size and shape of the footprints reveal the passerby to be a Mink.

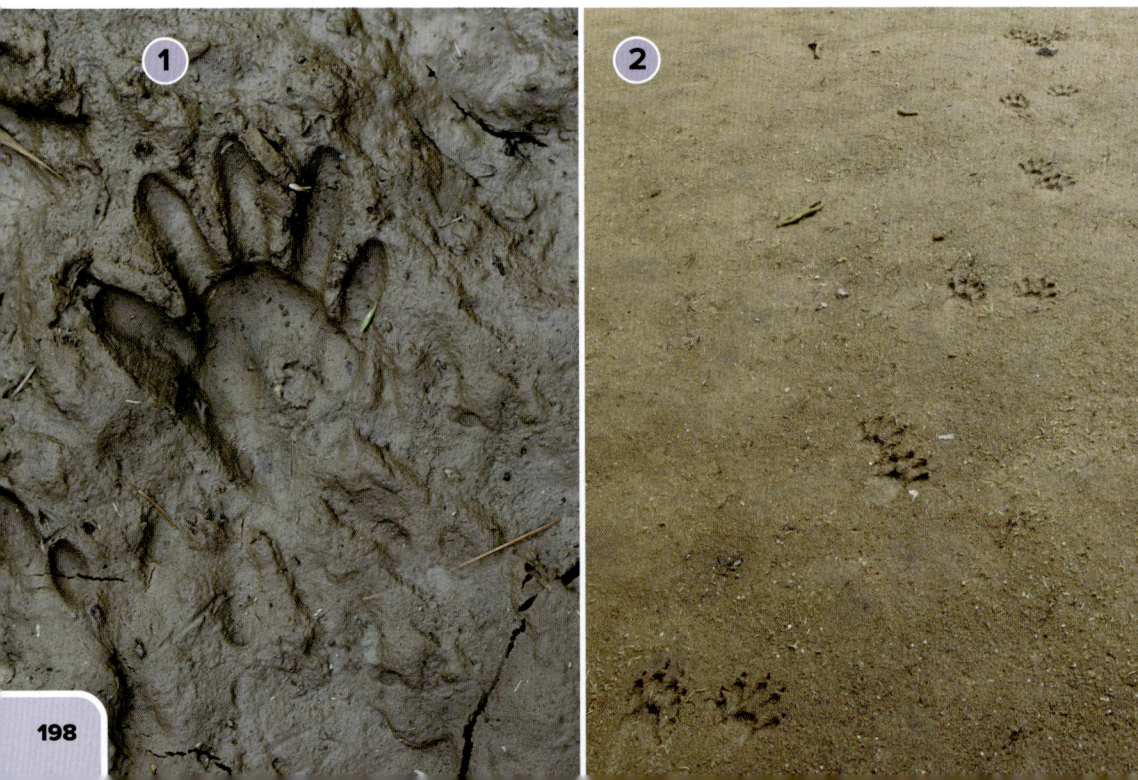

SIGNS OF ANIMALS

Raven Signs

1 If you see "whitewash" on a cliff face (like the three white patches on the top right of this cliff), you are likely seeing the nest and/or roost site of a pair of Common Ravens.

2 Looking more closely, one should eventually spot a bulky stick nest close to one of the whitewash patches. The upper patch of white is below the ledge of the pair roosts while the other attends the nest.

Wildflowers

Wildflowers are plants with non-woody (herbaceous) stems. In the Water includes species that are grow on or emerge from the water, while wildflowers On the Shore are species not found in the water. Within those categories, species are grouped by the color of their flowers.

WILDFLOWERS | IN THE WATER

White Water-lily

White Water-lilies often carpet the water's surface on slow-moving creeks and shallow lake bays. The flowers open in the morning and close at night, changing their sex over the first two days from pollen-receiving females to pollen-giving males.

Water Arum

Pretty to look at but nasty to eat, Water Arum is full of calcium oxalate crystals. The white spathe that embraces the flower spike (spadix) gives it a Calla Lily-like appearance, hence its other name — Wild Calla. This one has gone to seed.

WILDFLOWERS | IN THE WATER

Northern Water-plantain

With its tiny flowers and multi-branched thin stems, Northern Water-Plantain might appear to be a delicate plant, but it is actually quite hardy, able to survive drought better than many of its more robust aquatic neighbors.

Common Arrowhead

Arrowhead, named after the shape of its leaves, likes shallow water. All parts of the plant are reportedly edible, especially the corm, the swollen storage organ at the base of the stem.

WILDFLOWERS | IN THE WATER

European Frog-bit

In 1932 European Frog-bit was brought to the Central Experimental Farm in Ottawa to be considered for use as an ornamental plant. In 1939 this non-native species was found in the Rideau Canal and since then has invaded much of the Northeast.

IN THE WATER | **WILDFLOWERS**

Seven-angled Pipewort

1 Looking like anything but a wildflower, Pipewort can form impressive stands in shallow water.

2 Both male and female flowers are housed in one flowerhead, which is elevated above the water by a stalk whose length varies with water depth; the deeper it is, the longer the stalk.

WILDFLOWERS | IN THE WATER

Water Lobelia

The flower of a Water Lobelia sits up to a foot above the water, but the submerged portion of its hollow stem rises as much as a meter above the bottom!

IN THE WATER | **WILDFLOWERS**

Bog Bean

Not a bean but a wildflower of shallow fens and bogs, Bog Bean (also called Buckbean) has a very short flowering season. Fortunately, its name-giving leaves make it quite recognizable long before or after it blooms.

WILDFLOWERS | IN THE WATER

Little Floating-heart

The leaves of Little Floating-heart look like miniature versions of this wildflower's larger relative, the White Water-Lily. The Seminole used this plant to treat coughs and shortness of breath.

Sweetflag

At first glance, a colony of Sweetflag may look like a stand of cattail or Wild Iris, but its odd flower spikes soon reveal its true identity. When broken, Sweetflag leaves release a sweet smell that gives rise to its name. Prized by First Nations communities for its medicinal value, Sweetflag was a popular trade item, and its wide distribution in North America likely reflects this.

WILDFLOWERS | IN THE WATER

Yellow Pond-lily

Yellow Pond-lily, also known as Variegated Pond-lily or Bullhead Pond-lily, is a favorite food of Beavers, which eat the flowers, leaves and underwater rhizomes. The large, heart-shaped floating leaves provide perches for frogs, as well as dragonflies and damselflies. The Small-leaved Pond-lily looks like a small version of Yellow Pond-lily with a red central disk and leaves that are more deeply indented.

IN THE WATER | **WILDFLOWERS**

Beck's Beggarticks

Beck's Beggarticks (also known as Beck's Water-marigold) is the only member of its group that is fully aquatic. In addition to having leaves above the water, this wildflower also has finely dissected and feathery, submerged leaves.

WILDFLOWERS | IN THE WATER

Common Bladderwort

1 The tiny flowers of Common Bladderwort are the tip of the iceberg, for underwater lies a sprawling network of their stems, laden with tiny bladder-like leaf traps.

2 Bladderworts are "carnivorous" wildflowers, and a paddle often becomes heavy with a sprawling network of bladder traps.

3 The bladders contain a vacuum that sucks in water, along with the tiny creature that triggered the trap door to snap open. In the blink of an eye, the door shuts closed, trapping its victim. The prey is digested, and the water is eventually pumped back out, resetting the trap.

IN THE WATER | **WILDFLOWERS**

WILDFLOWERS | IN THE WATER

Watershield

1 Watershield is named after how the stems of the oval leaves are attached to the center of their underside. When you free a tangled mass of these plants from your paddle, you immediately feel a slippery coat of mucilage. Special hairs create this coating on the stems and underside of the leaves that is believed to protect those parts from insect attack. However, it offers no safety from Moose, which devour them for their sodium content.

2 Water-shield's tiny flowers are much less commonly noticed than its floating leaves.

IN THE WATER | **WILDFLOWERS**

Marsh Cinquefoil

Although Marsh Cinquefoil can have five leaflets in each compound leaf (as the name "Cinquefoil" would have you think), the number of leaflets can be as high as seven. Marsh-cinquefoil is the most aquatic member of its group.

WILDFLOWERS | IN THE WATER

Water Smartweed

1 Water Smartweed often emerges in large numbers from shallow water, where the current is weak.

2 This stunning wildflower can also grow in moist mud, an amphibious feat that gives rise to the second (species) part of its scientific name: *Persicaria amphibia*. In mud the plants develop trichomes, defensive hairs not found on plants growing in the water's safety.

Swamp Loosestrife

Swamp Loosestrife (also called Water Willow due to its willow-like leaves) is a native wildflower/shrub that can form dense colonies along slow-moving waterways. Swamp Loosestrife's flowers are similar to those of Purple Loosestrife (three forms, with only one form per plant). There is one difference in their reproduction; where Swamp Loosestrife's arching stems touch down, new plants arise by cloning.

WILDFLOWERS | IN THE WATER

Flowering Rush

Flowering Rush is a European plant that arrived in North America more than a century ago and continues to spread through Northeastern waterways. While it is undeniably a beautiful plant, it is nevertheless considered to be a highly invasive one.

Sphinx Ladies' Tresses

1 Sphinx Ladies' Tresses is a recently named species of orchid (formerly known as Nodding Ladies' Tresses) that blooms at summer's end, often along newly exposed muddy shores.

2 There are a number of species of Ladies' Tresses, all of which sport small white flowers typically growing in a spiral. The amount of yellow at the base of the labellum, the enlarged petal that serves as a landing strip for insect visitors, is one feature that helps separate the different species.

WILDFLOWERS | ON THE SHORE

Tall White Bog Orchid

Tall White Bog Orchid grows in fens where calcium is present. This distinctive orchid owns the peculiar trait of smelling like cloves.

ON THE SHORE | **WILDFLOWERS**

White Turtlehead

1 The odd flowers of White Turtlehead sometimes have a pink blush, and — especially before they open — do resemble a tortoise head.

2 Turtlehead is the primary food of Baltimore Checkerspot caterpillars, which become one of our most beautiful butterflies.

WILDFLOWERS | ON THE SHORE

Marsh Bellflower

You may have to scrutinize the edges of marshes to spot Marsh Bellflower, for it is small but nevertheless beautiful. It has a weak stem, so it sprawls over other vegetation.

Boneset

It is myth that Boneset helps set broken bones, but European settlers did use an infusion of this plant to treat constipation and colds. Many insects are attracted to its late summer blooms for nectar, but herbivores seldom eat Boneset leaves because, as they are loaded with chemical defences.

WILDFLOWERS | ON THE SHORE

American Cow-parsnip

American Cow-parsnip is a giant wildflower, growing two or more meters tall. It might be pleasing to the eye, but can give you an unpleasant experience if you make physical contact with one, for, like many other members of the Carrot family, it contains photo-active chemicals that make your skin sensitive to sunlight, resulting in burn-like blisters.

ON THE SHORE | **WILDFLOWERS**

Bulblet-bearing Water-hemlock

Don't let the flimsy appearance of a Bulblet-bearing Water-hemlock fool you — it is a deadly poisonous plant. These plants are found at the edge of marshes but no matter how hungry you are, DO NOT eat them!

WILDFLOWERS | ON THE SHORE

Spotted Water-hemlock

Spotted Water-hemlock is a toxic plant that, if eaten, gives humans and livestock very nasty side-effects — even death. Its toothed leaves are different from its relatives in that the veins end in the notches of the teeth, not in their pointed tips.

Water-parsnip

Water-parsnip, like other members of the Carrot family, has flowers held in clusters called umbels. An umbel is like an umbrella, with the stems of the floral heads radiating out from the main stalk. Unlike other members of its family, Water-parsnip roots are edible — but BEWARE of this plant's similarity to its poisonous relatives!

WILDFLOWERS | ON THE SHORE

Tall Meadow-rue

Growing as tall as a shrub, Tall Meadow-rue often cloaks shorelines in white in mid-summer. Its flowers are white but not because of petals; that color arises from its clubbed stamens that adorn male plants and stigmas that adorn female plants.

ON THE SHORE | **WILDFLOWERS**

Tall White Aster

An aster "flower" is actually a composite of two flower types: petal-like, ray flowers and disk flowers that form a central cone. If you pull out a ray, at its base you will see a tiny female flower. Disk flowers contain both sexes. Also known as Flat-topped Aster for its overall shape, each composite flower of Tall White Aster contains 4 to 15 petal-like ray flowers.

WILDFLOWERS | ON THE SHORE

White Panicled Aster

The composite flowers of White Panicled Aster, a late-bloomer that decorates shorelines, each have 16 to 47 white petal-like rays.

ON THE SHORE | **WILDFLOWERS**

Rose Pogonia and Grass Pink

1 In early summer, Rose Pogonias carpet boggy shorelines. These beautiful orchids lure pollinators with false promises of food, achieved through modified hairs that look like pollen-bearing stamens.

2 Look among Rose Pogonias for Grass Pinks. The enlarged petal (labellum) of this multiple-bloomed orchid is the uppermost, the only orchid with this feature. Its pseudo-stamens deceive insects looking for pollen into landing on them.

WILDFLOWERS | ON THE SHORE

Dock

Docks are a large group of similar-looking species, many non-native, which have been described as "coarse, homely plants." Several grow on shores and in wetlands, towering above other plants, and when their fruit ripens, they add spikes of color to their habitat.

Cardinal Flower

1 Few wildflowers stand out from their background more strongly than a Cardinal Flower. These stunning plants prefer calcium-rich soil beside flowing water.

2 The vivid red blooms are invisible to insects, but not to Ruby-throated Hummingbirds, their only known pollinator. As a hummingbird extracts nectar from a Cardinal Flower's long spur, its head rubs against the pollen-bearing stamens. After the stamens have served their purpose, they get replaced by the female sex organ, the stigma, which receives pollen from future visitors.

WILDFLOWERS | ON THE SHORE

Pitcher-plant

1. Pitcher-plant is a "carnivorous" plant that steals nitrogen from the bodies of insects captured by its modified leaves, which serve as pitfall traps. Insects walk down into the pitcher where downward-pointing hairs discourage an exit. Suddenly the visitor reaches a curved, slippery slope and down it slides down it into rainwater held by the pitcher, its fate now sealed.

2. A Pitcher-plant's flowers are as peculiar as its leaves.

ON THE SHORE | WILDFLOWERS

Swamp Milkweed

As in other Milkweeds, the nectar of Swamp Milkweed is held in little cups for all to enjoy, but with one caveat: the patron must have the strength to pull its leg free from a slit where a saddlebag of pollen clamped onto it. If successful, when the burdened foot slips down a slit in another flower from which pollen was removed, the saddlebag unclamps, and pollination is facilitated.

WILDFLOWERS | ON THE SHORE

Spoon-leaved Sundew

1 Spoon-leaved Sundew and its relatives paint logs and muddy shores in the North a soft pink-red hue. These "carnivorous" plants thrive in nutrient-poor areas by gleaning nitrogen from insects they capture with their adhesive leaf traps.

2 A Sundew's finger nail-size leaf (here, Spatulate-leaved) is equipped with gland-tipped hairs. The outer hairs release glue that sticks to whatever lands on them while the inner, shorter hairs release enzymes that digest the prey.

ON THE SHORE | **WILDFLOWERS**

Round-leaved Sundew

Once an insect is captured, the leaf of a Sundew (here, Round-leaved) slowly folds over, pressing the victim against the enzyme-releasing hairs in a lethal embrace. Large prey like this Chalk-fronted Corporal may require a collective effort to deal with it!

WILDFLOWERS | ON THE SHORE

Spotted Joe-Pye Weed

"Weed" has the negative connotations of alien, invasive, and ugly, so being a beautiful and common native plant, Spotted Joe-Pye Weed is a misnomer. Its late summer blooms attract a diverse array of insect visitors, including butterflies.

ON THE SHORE | **WILDFLOWERS**

Small Purple Fringed Orchid

If you spot the reclusive Small Purple Fringed Orchid peering through the shoreline vegetation, be sure to pull up for a closer view. The petals of this spectacular orchid (which are often pink) gave rise to the species part of its scientific name *Platanthera psycodes*: "psycodes" is Greek for butterfly.

WILDFLOWERS | ON THE SHORE

Wild Iris

Wild Iris, also called Blue Flag, is a stunning but complex wildflower. Each flower consists of three colorful petal-like sepals and three shorter true petals placed above them. Its main pollinators are Bumble Bees that must muscle their way between a petal and a sepal to access nectar at the base of a sepal. As it enters, the bee's back rubs against the flower's sexual parts, facilitating pollination.

Bog Aster

Bog Asters appear in late summer on wet shorelines as well as floating peatlands. Each plant has small, crowded leaves and bears at best only a few pink flower heads.

WILDFLOWERS | ON THE SHORE

Purple-stemmed Aster

Also known as Swamp Aster, Purple-stemmed Aster is a distinctive species, having between 30 and 50 purple "ray flowers," and leaves that hug a purple stem covered in stiff white hairs.

ON THE SHORE | **WILDFLOWERS**

Purple Loosestrife

1 Once considered a threatening invasive species, Purple Loosestrife seems to be taking a lesser role in wetlands, and has even been found to benefit some native plants by attracting large numbers of pollinators.

2 Like Swamp Loosestrife and Pickerelweed, a plant bears only one of three flower forms, each differentiated by the length of the pistil's style ("stem") in relation to two sets of pollen-bearing stamens. This is a short-style form, its green stigma slightly visible in the right flower.

WILDFLOWERS | ON THE SHORE

Blue Vervain

Blue Vervain's blue/purple flowers appeal not only to our eyes. Bees selectively visit flowers of this color, so Vervain attracts a steady stream of pollinators. Ironically, a potion made by the Iroquois from this beautiful shoreline wildflower apparently made "obnoxious persons go away."

ON THE SHORE | **WILDFLOWERS**

Square-stemmed Monkeyflower

The flowers of Square-stemmed Monkeyflowers are said to resemble a monkey's face, hence its name. A study revealed that if Purple Loosestrife grows near this plant, the pollinating Bumble Bees choose the blooms of that non-native Loosestrife over the native monkeyflower.

WILDFLOWERS | ON THE SHORE

Marsh Skullcap

Marsh Skullcap hides very well in other vegetation, so it may take a bit of looking to see one, even though it is a common shoreline plant. The name comes from its upper petal, which is reputed to look like a skull.

ON THE SHORE | **WILDFLOWERS**

Swamp Candles

1 A stunning wildflower that often forms large colonies, Swamp Candles, also known as Yellow Loosestrife, is native to eastern North America but was introduced in the Northwest, where it is considered invasive in cranberry bogs and blueberry operations.

2 Swamp Candle is an attractive wildflower, especially to the oil bees (Melittids), which go to its flowers to collect a special oil, one used by very few plants to attract pollinators.

WILDFLOWERS | ON THE SHORE

Tufted Yellow Loosestrife

Female Melittid bees use special hairs on their front and hind legs to gather oil from the flowers of Tufted Yellow Loosestrife. The oil is used to waterproof their nest cell linings and is mixed with pollen to feed young bees. A poultice made from Tufted Yellow Loosestrife was used by the Iroquois to stop milk flow in mothers who finished nursing their babies.

ON THE SHORE | **WILDFLOWERS**

Horned Bladderwort

Most Bladderworts "hunt" in water, but Horned Bladderwort does so on mucky shores, its bladder traps hidden in mud. Their trap door has been timed opening in a few milliseconds, making it one of the fastest-moving floral parts in the world!

WILDFLOWERS | ON THE SHORE

Nodding Beggarticks

1 Nodding Beggarticks often form large colonies along shores. Beggarticks are named for their seeds, which have barbs for hitching rides on animals.

2 Like other members of the Aster family (Asteraceae), a Nodding Beggartick "flower" is actually a composite of petal-like ray and densely crowded disk flowers.

ON THE SHORE | **WILDFLOWERS**

Grass-leaved Goldenrod

While most goldenrods prefer dry habitats, Grass-leaved Goldenrod often grows in moist ground. Its leaves are much narrower than most members of its group.

WILDFLOWERS | ON THE SHORE

Fall Sneezeweed

1 Despite its name, Fall Sneezeweed does not cause allergies. Rather, a powder derived from dried plants induces sneezing, which is why Indigenous Americans used Sneezeweed to treat congestion and headaches.

2 Two distinctive features of Sneezeweed's ray petals are their droop and their end in three "teeth."

ON THE SHORE | **WILDFLOWERS**

Jewelweed

A Jewelweed's flower first opens as a male that gets its pollen removed by bumble bees and hummingbirds. Eventually the male structure falls off, revealing a green pistil (top left flower), a sex change that prevents self-pollination. When Jewelweed seedpods mature (bottom right), the slightest touch causes them to audibly snap open, throwing the seeds several meters from the plant, giving rise to this wildflower's other name: Spotted Touch-me-not!

Woody Plants

Unlike Wildflowers, these plants have woody stems. Subcategories are Shrubs, Trees and Vines. Shrubs are smaller than trees, reaching a maximum height of two to three meters, and many bear showy flowers, a feature that has some species included in wildflower books. Vines crawl up tree trunks or sprawl atop other plants.

Eastern White Cedar

1 Eastern White Cedars are common shoreline residents. When cedars lining a shore appear pruned at the bottom, it is because, in winter, White-tailed Deer devour the lower foliage, creating a distinct "browse line."

2 Cedar cones are small, but in some years, they are abundant, providing food for Red Squirrels, Black-capped Chickadees and small finches such as Pine Siskins.

TREES | **WOODY PLANTS**

WOODY PLANTS | TREES

Eastern White Pine

1 With their towering stature and sweeping branches, Eastern White Pines add elegance to shorelines.

2 All pines have needles grouped in bundles; the long, soft needles of Eastern White Pine are grouped in bundles of five.

3 White Pines, like other trees, have years of low seed production, then unpredictably produce a mast crop. This overwhelms the seed predators such as the Red Crossbill seen atop this tree, allowing some seeds to avoid consumption and germinate.

TREES | WOODY PLANTS

WOODY PLANTS | TREES

Red Pine

1 The reddish bark and "clumpy" appearance of a Red Pine is distinctive.

2 Red Pine needles are long and stiff and grouped in bundles of two. Their cones are held tight to a branch, unlike the dangling cones of Eastern White Pines.

TREES | **WOODY PLANTS**

Jack Pine

1 The tree immortalized by Tom Thomson, Jack Pine is a northerner with a gnarly appearance.

2 Jack Pine needles are short and stiff and arranged in clusters of two. A temperature of at least 50°C is required to melt the resin that holds their cone scales closed. Fires generate that heat, and due to a delayed opening of the scales, the seeds fall out unscathed and germinate in the mineral soil below.

WOODY PLANTS | TREES

Tamarack

1 Also known as Larch, the Tamarack is the only native coniferous tree that loses its needles in the fall (making it a deciduous coniferous tree). Its soft needles are arranged in tufts of 10 to 20, and their female flowers look like little roses.

2 Before Tamarack's needles are shed in November, they turn brilliant gold, adding a stunning element of beauty to an otherwise drab month.

TREES | **WOODY PLANTS**

Balsam Fir

1 Balsam Fir boreal forest are trees that usually have a spire-shaped crown.

2 The flat needles are usually positioned in a single row on both sides of the twigs. Balsam Fir is the only coniferous tree with its cones positioned upright.

WOODY PLANTS | TREES

White Spruce

1 White Spruce grow on drier sites than Black Spruce does, and have a "fuller" crown and longer branches that grow out from most of the trunk. Unlike those of Balsam Fir, spruce cones hang down.

2 Spruce needles are individually attached to twigs (not grouped in clusters like pine needles) and are rounder than the flat needles of Balsam Fir. White Spruce needles are longer and more aromatic than those of Black Spruce.

TREES | **WOODY PLANTS**

Black Spruce

Black Spruce are also boreal forest trees that dominate peatland habitats. Their growth form is typically "spindly," with tall, relatively narrow trunks sporting a clump of short branches near the top.

WOODY PLANTS | TREES

Eastern Hemlock

1 Eastern Hemlocks often grow by the water's edge on north-facing slopes and can be as tall as a pine. Overall, a hemlock's shape looks a bit untidy, lacking the elegant sweeping branches of an Eastern White Pine.

2 At close view, a hemlock's short flat needles become obvious, as do its tiny cones — an inch or so long — which might seem undersized for such a huge tree!

TREES | **WOODY PLANTS**

Sugar Maple

1 Sugar Maple leaves have smooth margins between their five main points. If you spot a flash of red high up in the maple canopy, you are likely seeing a Scarlet Tanager, a bird that sounds like a robin with a sore throat!

2 Sugar Maple trees provide us with more than just maple syrup; in the fall, they put on a visual extravaganza, one that is unrivalled.

WOODY PLANTS | TREES

Red Maple

1 Red Maple leaves contain small indentations between their main points. They look a bit like Mountain Maple leaves, but Red Maple leaves have a "glossy" finish, not "matte" like those of the Mountain. Sugar Maple leaves lack small indentations between the main points.

2 Maples typically bear both male and female flowers on the same tree but most Red Maples bear only male or female flowers. Amazingly, in the autumn, the leaves of male Red Maples turn red while those on the female trees turn yellow!

TREES | **WOODY PLANTS**

Silver Maple

1 Silver Maples like to have their feet wet, so they often grow in river floodplains.

2 Their leaves are more deeply cut than those of their relatives, but like all maples, they produce samaras (keys): winged seeds that spin like little helicopters, carrying the seeds some distance from the mother tree.

WOODY PLANTS | TREES

Black Ash

Decimated by the invasive Emerald Ash Borer in recent years, large Black Ashes are now sadly few and far between. Ashes have compound leaves (one leaf divided into numerous leaflets), with Black having as many as 11 stemless leaflets attached directly to each twig. The seeds (achenes) of Black Ash, like other ashes, have a single wing for dispersal.

TREES | **WOODY PLANTS**

Showy Mountain-ash

Not a true ash, but a member of the Rose family, Showy Mountain-ash produces fruit that birds such as American Robins and Cedar Waxwings quickly devour. There are two native species in the Northeast. Showy Mountain-ash has relatively broad compound leaves that end in an abrupt point; those of the similar American Mountain-ash are narrower with longer, tapering tips.

WOODY PLANTS | TREES

American Mountain-ash

One never realizes how common Mountain-ash is along shorelines until it bears fruit, or its leaves turn color in the fall. American Mountain-ash have narrower leaves with longer tapering tips than those borne by Showy Mountain-ash.

TREES | **WOODY PLANTS**

Basswood

1 Basswood have large heart-shaped leaves with strongly serrated edges. They grow quickly and tall and often develop large cavities that provide homes for Raccoons, Porcupines and Barred Owls.

2 A Basswood's pleasant-smelling flowers are popular with bees, and when the fruits (nutlets) mature, a snowboard-shaped bract serves as a helicopter blade to carry them some distance from the mother tree.

WOODY PLANTS | TREES

White Birch

1 White birch are known for their white, peeling, paper-like bark that is used by Indigenous Peoples to build canoes.

2 In spring, the long yellow catkins of White Birch are as pretty as any spring wildflower.

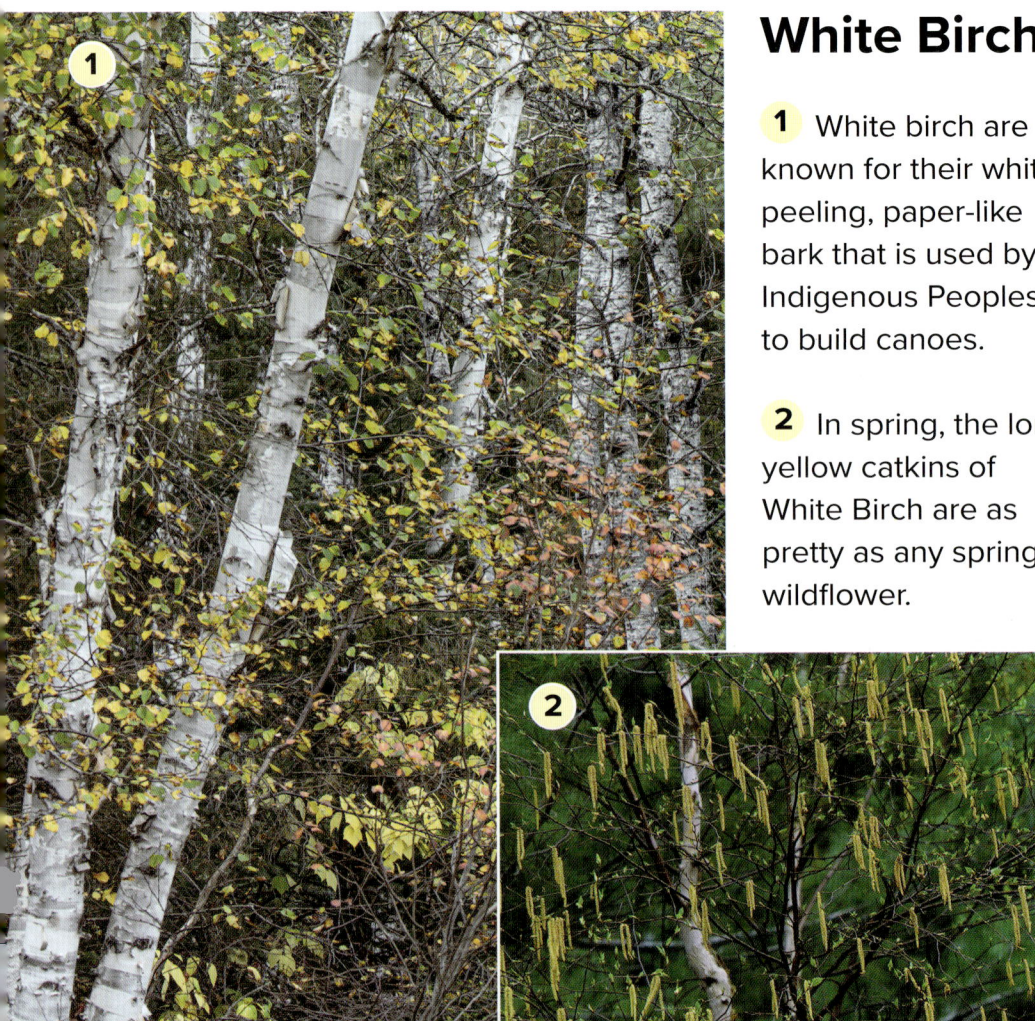

TREES | **WOODY PLANTS**

Balsam Poplar

Balsam poplars have shiny, longish leaves often with orange resin on their underside. The buds also contain this resin, which gives what is likely the most distinctive feature of this tree: its sweet smell.

WOODY PLANTS | TREES

Trembling Aspen

1 Also known as Poplar, Trembling Aspen is a sun-loving tree with a smooth, pale trunk, often colonizing areas burned by fire.

2 Its leaves have a long, flattened stem, which makes them "tremble" under the slightest breeze. The flowers appear as long catkins.

3 Trembling Aspen bark is a favorite food of Beavers, and its spring flower buds are a staple for Ruffed Grouse and Black Bears.

TREES | **WOODY PLANTS**

WOODY PLANTS | TREES

Cottonwood

1 Although not found through most of the Northeast, Cottonwood can be common growing in sandy areas (especially dunes) along Lake Ontario and Lake Erie. Older trees have deeply furrowed bark.

2 Cottonwood leaves are strongly triangular with numerous teeth on the sides. As for other poplars, the seeds are adorned with a parachute of long silky hairs that look like cotton.

TREES | **WOODY PLANTS**

Pin Cherry

Pin Cherry has narrower leaves than Choke Cherry, and its berries (and flowers) are held on longer stalks and in much smaller clusters.

WOODY PLANTS | TREES

Choke Cherry

1 In late spring, the long clusters (racemes) of Choke Cherry flowers make it an easy tree or shrub to identify.

2 The clusters of berries do not last long for they are devoured by Black Bears, Cedar Waxwings, and American Robins, to name a few! Eastern Chipmunks harvest them for the winter, eating their pits every few days when they awaken from their slumber.

SHRUBS | **WOODY PLANTS**

Shrubby Cinquefoil

1 The compound leaves of Shrubby Cinquefoil often contain five leaflets, hence "cinquefoil." **2** The beautiful yellow flowers are the reason this shrub (or a cultivar of it) is commonly used as a garden ornamental.

WOODY PLANTS | SHRUBS

Smooth Serviceberry

The flowers of Serviceberries add sprays of white to the landscape just as most trees are starting to leaf out. The bronze half-open leaves at flowering time, reddish twigs and young branches, and flowers borne on long stalks reveal this to be Smooth Serviceberry. While some species remain shrubs, others, including Smooth Serviceberry attain the status of small trees.

SHRUBS | **WOODY PLANTS**

Black Chokeberry

1 Black Chokeberry grows along peatland and marsh edges. If you look closely at the right leaf, you will see the tiny dark hair-like glands along its midrib. The edges of the leaves bear these as well.

2 The fruit of Black Chokeberry may look tasty, but due to a high tannin content it is highly astringent and makes your mouth pucker if you dare taste one!

WOODY PLANTS | SHRUBS

Ninebark

1 Along with peeling bark, Ninebark's three-lobed leaves and hemispherical flower heads make it an easy shoreline shrub to identify.

2 Ninebark's seedpods arguably are as attractive as its flowers.

SHRUBS | **WOODY PLANTS**

Red-osier Dogwood

1 Red-osier Dogwood often forms dense thickets along waterways. Its red branches are especially conspicuous after the leaves have fallen.

2 Many dogwoods have white or pale blue fruit (technically a drupe), which has within the soft pulp a seed encased in a stony coated.

WOODY PLANTS | SHRUBS

Gray Dogwood

Obviously not named for the color of its fruit stalks, Gray Dogwood sometimes forms thickets along waterways in areas with basic pH (i.e., limestone bedrock), not on the acidic Canadian Shield.

SHRUBS | **WOODY PLANTS**

Leatherleaf

1 Leatherleaf often forms dense carpets on floating peatlands and boggy shores providing ideal perches for Crimson-ringed Whitefaces. As in other Ericaceous plants, Leatherleaf's leaves persist through winter to conserve nutrients in the shrub's nutrient-poor habitat.

2 Leatherleaf's dainty flowers bloom early, not long after the ice leaves northern waterways.

WOODY PLANTS | SHRUBS

Labrador Tea

1 A common shrub in the North, Labrador Tea tends to grow on older peatlands, especially under partial shade of Black Spruce.

2 The thick evergreen leaves are retained all winter to conserve nutrients, and the dense hairs on their underside help prevent desiccation.

SHRUBS | **WOODY PLANTS**

Buttonbush

Each ball-like flower cluster of a Buttonbush contains up to 200 individual flowers. The "halo" around the clusters is created by the long, thin styles of the female sex organs. The distinctive flower heads in July are replaced by spherical clusters of cone-shaped nutlets in September.

WOODY PLANTS | SHRUBS

Mountain Holly

1 Mountain Holly (right) and Winterberry (left) are both native species of holly that grow along the water's edge.

2 How their berries (technically, drupes) attach to the branches readily help identify the two. Those of Mountain Holly have long stems, while those of Winterberry have none.

SHRUBS | **WOODY PLANTS**

Winterberry

The attractive berries (drupes) of Winterberry lack stems and are held tight to the twigs. The berries persist through the autumn and liven winter landscapes, giving rise to this shrub's name.

WOODY PLANTS | SHRUBS

Wild Raisin

1 Wild Raisin is a common shoreline shrub on the Canadian Shield. Its opposite-positioned pairs of leaves bear a distinct pale central line.

2 The colorful fruit never lasts long for it is devoured by a great variety of birds including warblers such as this young Yellow-rumped.

SHRUBS | WOODY PLANTS

Highbush Cranberry

1 Highbush Cranberry has distinctive flowers that are of two types: inner ones that bear sexual parts; and large outer sterile flowers. It is speculated that the outer ones serve to attract pollinators.

2 Birds devour Highbush Cranberry berries (and disperse their seeds), ensuring that the attractive fruit doesn't linger long on the shrubs. A near-identical European variety is commonly planted as an ornamental, but its fruit seems scorned by birds and remains on the shrubs all winter.

WOODY PLANTS | SHRUBS

American Bladdernut

American Bladdernut grows along river banks and rocky slopes as far north as southern Ontario and northern Vermont. Its compound leaves are distinctive, with three leaflets that are sharply serrated. The name arises from the odd bladder-like pods that encase the seeds.

SHRUBS | **WOODY PLANTS**

White Meadowsweet

1 White Meadowsweet is the foodplant for the Spring Azure caterpillar. There are two varieties of White Meadowsweet. "Broad-leaved" Meadowsweet has wider leaves and a broader cluster of flower heads than "Narrow-leaved" Meadowsweet.

2 In addition to having more slender leaves, "Narrow-leaved" Meadowsweet has a tight pyramidal shape.

WOODY PLANTS | SHRUBS

Rosy Meadowsweet

The pink spire-like flower heads of Rosy Meadowsweet (also called Steeplebush) make it an easy wildflower to identify.

SHRUBS | WOODY PLANTS

Sheep Laurel

Sheep Laurel, a common shrub of peaty shores, is worthy of a closer look. The stamens are pinned back in grooves until their arm (filament) is touched by a pollinator. The stamen suddenly springs free, dashing pollen onto the undoubtedly startled intruder. Sheep Laurel blooms in summer and, unlike its nearest relative, Bog Laurel, has leaves above its flowers as well as below them.

WOODY PLANTS | SHRUBS

Bog Laurel

Bog Laurel (on the right with Bog Rosemary on the left) blooms in spring much earlier than its larger cousin, Sheep Laurel, and lacks leaves above its flowers. Bog Rosemary leaves may look like those of a popular savory herb but unfortunately, they taste nothing like them!

SHRUBS | **WOODY PLANTS**

Prickly Wild Rose

Prickly Wild Rose (also called Bristly Rose) grows on shores and river banks. While the flowers of other wild Roses are identical, the stipules (leaf-like appendages at the stem base) and stem prickles vary in appearance between species. If you see a rose growing in a boggy swamp, it is likely Swamp Rose, which has fewer but larger downturned prickles.

Speckled Alder

Anyone who has paddled narrow northern waterways is familiar with Speckled Alders, named for the white marks (lenticels) on the stems and branches. Alder roots contain orange root nodules full of symbiotic bacteria. The bacteria transform ("fix") nitrogen into a form usable by the shrubs, which in turn provide sugars for the bacteria.

SHRUBS | **WOODY PLANTS**

Green Alder

Green Alder are best told from Speckled Alder by their female flowers and cones that dangle from the stems on long stalks; those of Speckled lack stalks.

WOODY PLANTS | SHRUBS

Dwarf Birch

Also known as Bog or Swamp Birch, Dwarf Birch frequents shorelines and wetlands with underlying limestone. Its tiny leaves are heavily toothed and almost round.

SHRUBS | **WOODY PLANTS**

Bebb's Willow

1 Willows can be a difficult group of species to identify. Also known as Long-beaked Willow, Bebb's Willow is the most common and widespread willow to reach tree-size.

2 Bebb's Willow is a favorite food of Moose, and Beavers clearly take advantage of its clustered growth form.

WOODY PLANTS | SHRUBS

Shining Willow

One of the more distinctive willows, Shining Willow has glossy leaves with finely serrated edges. It reaches the stature of a small tree, and it is the tallest native willow in Newfoundland and Labrador.

SHRUBS | **WOODY PLANTS**

Bog Willow

Bog Willow is a peatland inhabitant, growing to about a meter in height. Its smooth-margined leaves are shorter and more oval than those of other willows. The red-brown female flowers provide a lovely contrast to the leaves.

WOODY PLANTS | SHRUBS

Sweet Gale

1 A northern shrub with nitrogen-fixing nodules on its roots, Sweet Gale is easily identified by crushing and smelling its aromatic leaves, which were historically used to aromatise linen and flavour beer.

2 If you are paddling soon after the ice leaves the northern waterways, Sweet Gale's yellow-brown male catkins dominate many a shoreline. The inconspicuous female flowers are carried on separate plants.

SHRUBS | **WOODY PLANTS**

Glossy Buckthorn

Glossy Buckthorn was brought to North America from Europe as an ornamental shrub more than a century ago. Despite its name, it has no thorns. Glossy Buckthorn is highly invasive, forming dense stands that crowd out native species. Each berry contains three to four seeds that can lie dormant for several years before germinating.

WOODY PLANTS | VINES

Virgin's Bower

1 Virgin's Bower is a vine that sprawls on top of other vegetation so that its leaves can reach sunlight.

2 At summer's end and through the fall, the distinctive pom-pom seedheads of Virgin's Bower's reveal just how common this vine can be.

VINES | **WOODY PLANTS**

Virginia Creeper

Virginia Creeper is most visible in the autumn, when the vine's leaves turn red, and its ripe fruit turns purple.

WOODY PLANTS | VINES

Riverbank Grape

The fruit of Riverbank Grape is enjoyed by many birds, including Northern Cardinals and Pileated Woodpeckers.

VINES | **WOODY PLANTS**

American Bittersweet

1 American Bittersweet is a native vine that grows on top of other shrubs and trees or sprawls across rocky slopes. The flower sexes are separate, with individual plants producing either all male or all-female flowers.

2 American Bittersweet fruit is initially round with an outer casing that splits open in the fall to reveal the bright red, berry-like, seed-containing aril housed inside. The non-native invasive Oriental Bittersweet sold by nurseries looks similar but has broad leaves that lack fine serrations, and has flowers and fruit that are positioned along the stems, not just at their terminal end as in the native species.

Other Plants

This chapter features non-flowering plants such as Ferns and flowering plants that lack showy blooms, notably Sedges and Grasses. Also included are Rushes, Pondweeds, Bur-reeds and Cattails. Lichens are not plants but symbiotic associations between fungi and algae or cyanobacteria but for convenience they are arbitrarily included in this category.

OTHER PLANTS | FERNS

Water Horsetail

1 Horsetails are spore-producing "living fossils" whose ancestors grew as tall as trees. Most species grow on land, but River Horsetails are aquatic, often forming large colonies.

2 Horsetail stems are jointed, and are tough due to high silica content. The toughness was exploited by pioneers who used horsetails as scouring pads. Water Horsetail stems are hollow to allow air transport to the plant's submerged parts.

FERNS | **OTHER PLANTS**

Sensitive Fern

Sensitive Fern love to have their feet wet. The fern's spore cases are held on a separate stalk that persists throughout winter. However, the sterile frond is not as hardy and succumbs to drought or frost, giving rise to the descriptive "sensitive."

OTHER PLANTS | FERNS

Royal Fern

The fronds of Royal Fern look like the compound leaves of a flowering plant, for they are divided in a similar way. However, the warm brown, spore-bearing spike that rises above the plant like a crown leaves no doubt as to it being a fern.

FERNS | **OTHER PLANTS**

Cinnamon Fern

The tall spikes of cinnamon-colored spore cases and large fronds make Cinnamon Fern distinctive. However, if there is no fertile spore-bearing spike, the sterile green fronds look just like those of Interrupted Fern, a relative that bears its spore cases in the middle of a regular frond, and not on a separate stalk.

OTHER PLANTS | FERNS

Marsh Fern

Marsh Ferns venture farther into wet habitats than most of their relatives do. They are commonly found in cattail marshes, on sedge mounds in fens, and along shores.

FERNS | **OTHER PLANTS**

Ostrich Fern

The large, arching fronds of Ostrich Ferns resemble an Ostrich feather and hence the name. This beautiful fern loves floodplains and sometimes covers banks along smaller waterways. When Ostrich Ferns start to rise from the ground in early spring, their coiled, highly edible "fiddleheads" are often over-enthusiastically collected.

Rusty Cliff Fern

1 Woodsias are small cliff-loving ferns and Rusty Cliff Fern (also called Rusty Woodsia) is a common species that grows on acidic cliff faces.

2 Rusty Cliff Fern has a lot of hairs, especially on the underside of its fronds where its sori (spore-containing "fruit dots") are located. By fall, the color of a frond's underside changes from white to rust.

FERNS | **OTHER PLANTS**

Fragrant Wood Fern

Named for its sweet-smelling glands, Fragrant Wood Fern is another rock-lover. Its shape is distinctive, tapered at both ends. The clump of dead fronds hanging below its base is also a good field mark.

OTHER PLANTS | FERNS

Marginal Wood Fern

1 Marginal Wood Fern is a leathery, evergreen fern that grows on rocky hillsides and outcrops, and is particularly common on the Canadian Shield. It is a "twice-cut" fern, with the fronds divided once from the main stalk into pinnae, and then those divided again into pinnules.

2 "Marginal" refers to the position of the spore-containing structures called sori. These are housed in clusters ("fruit dots") on the margins or edges of the underside of the pinnae.

FERNS | **OTHER PLANTS**

Rock Polypody

Rock Polypody is a rock lover that often keeps company with Marginal Wood Fern. It is a once-cut fern (the frond is dissected only once to its mid-rib). "Poly" means "many" and "podium" refers to "foot," which is a great description for this common fern.

OTHER PLANTS | FERNS

Maidenhair Spleenwort

Maidenhair Spleenwort is a distinctive, elegant little fern that grows on cool, shaded cliffs and talus slopes. It has two subspecies: one that grows on acidic rock such as granite, and another that prefers more basic rock, such as dolostone or limestone.

SEDGES | **OTHER PLANTS**

Canada Rush

1 Rushes superficially look like sedges but unlike sedges have tiny bisexual flowers clustered in a flowering head (inflorescence) and have round stems. Some species lack leaves while others have leaves that look like stems.

2 Many Rushes are reportedly wind-pollinated and so have no edible bribe to offer insects, so either this Canada Rush is insect-pollinated or the Flower Fly is out of luck!

OTHER PLANTS | SEDGES

Bayonet Rush

Even when not flowering, Bayonet Rush is identifiable, for it grows fairly tall in waters that fluctuate in depth, and shows red bases to the stems and stem-like leaves when the water recedes.

SEDGES | **OTHER PLANTS**

Lakeside Sedge

Lakeside Sedge belongs to the Carex sedges, which differ from other flowering plants in having their female flowers housed inside a sac-like structure called a perigynium, which are stacked together in spikelets. A close examination of perigynia is often required to identify a sedge's species.

OTHER PLANTS | SEDGES

Bearded Sedge

Especially attractive are sedges with large, inflated perigynia and "teeth." Bearded Sedge is a beauty with greatly divergent teeth that give it a "bottle brush" look.

SEDGES | **OTHER PLANTS**

Tussock Sedge

1 Sedges dominate wetlands in the same way that grasses dominate grasslands. Tussock Sedge is an important hummock-forming species and is a dominant species in sedge meadows.

2 Hummocks formed by Tussock Sedges are especially apparent in wetlands when water levels drop.

OTHER PLANTS | SEDGES

Lesser Bladder Sedge

Lesser Bladder Sedge is another common wetland sedge that grows in dense tufts.

SEDGES | **OTHER PLANTS**

Fringed Sedge

The slender dangling spikelets of Fringed Sedge are common sights along the edges of waterways. Like other members of the Carex genus, Fringed Sedge has triangular stems, giving rise to the saying: "Sedges have edges."

OTHER PLANTS | SEDGES

Northeastern Sedge

1 Often small sedges hide among shoreline vegetation and are easily overlooked. Northeastern Sedge is common where acidic conditions prevail.

2 Upon close examination, the inherent beauty of Northeastern Sedge becomes apparent. The brown staminate spike contains male flowers while the perigynia contain the female flowers.

Blunt Spikerush

Spikerushes are usually overlooked because they are small, simple-structured plants that emerge from shallow water or nearby mud. With no apparent leaves or flowers, Spikerushes have been described as "green matchsticks!" They can be difficult to identify, with Blunt Spikerush being a somewhat easier species to recognize.

OTHER PLANTS | SEDGES

Tussock Cottongrass

Cottongrass is a misnomer, for these delightful plants are actually sedges! The distinctive cottony appearance is due to numerous thread-like floral parts attached to each achene (one-seeded fruit). The number of spikelets and time of flowering helps separate species; Tussock Cottongrass has but one spikelet and blooms in early summer.

SEDGES | **OTHER PLANTS**

Tall Cottongrass and Tawny Cottongrass

1 Some species of Cottongrass have several spikelets per plant, and Tall Cottongrass is one of these. This northern species appears in late spring and early summer.

2 Tawny Cottongrass is the last of its group to make an appearance each year, its dense clusters of tawny spikelets decorating many a peatland in early autumn.

OTHER PLANTS | SEDGES

Swaying Bulrush

1 Drifting over a patch of Swaying Bulrush is mesmerizing for the submerged stems and leaves of this aquatic sedge gently sway in the current like a mermaid's hair.

2 Where the current slows, upright stems each bear a single spikelet of its tiny flowers.

SEDGES | **OTHER PLANTS**

Woolgrass

Woolgrass is actually not a grass but a sedge, which is why it is also called Woolsedge. In late summer, the huge, fuzzy flower heads of this plant tower along damp shores, nodding heavily in the slightest of breezes.

OTHER PLANTS | GRASSES

Canada Bluejoint

Grasses are notoriously difficult to identify. Most species grow in relatively dry habitats (such as grasslands), but Canada Bluejoint (Reed Grass) is a common wetland species. Initially its flowering heads are pyramid-shaped, but after flowering they close, forming tighter plumes that typically bend to one side.

GRASSES | **OTHER PLANTS**

Canada Manna Grass

The Manna Grasses are a pretty group, with Canada Manna Grass being a common wetland inhabitant.

OTHER PLANTS | GRASSES

European Common Reed

There are two different species of Phragmites: The highly invasive European Common Reed (here); and the native, non-invasive American Common Reed. They look extremely similar, with one difference being that the native species has red upper stem sections while the European species has tan ones.

GRASSES | **OTHER PLANTS**

Northern Wild Rice

1 Northern Wild Rice is a tall aquatic grass famous for its delicious grains, which are savored not only by our species!

2 As in other species pollinated by the wind, Wild Rice's female flowers are positioned above the male flowers (seen here) to allow the wind to carry the pollen to other Wild Rice plants, thereby avoiding self-pollination.

OTHER PLANTS

Pondweed

1 Pondweeds are common aquatic plants, with most species having thick, floating leaves in addition to thinner, submersed ones. However, they are a challenging group to identify to by species, usually requiring close scrutiny of leaf and stem details.

2 While the flowers of most Pondweeds are held above water and are wind pollinated, a few entirely submerged species have their pollen delivered by moving water, a rare type of pollination known as hydrophily.

OTHER PLANTS

Greater Duckweed

One of the smallest flowering plants in the world (although rarely seen in flower), Greater Duckweed floats on the surface of calm, shallow waters. In the fall, "buds" called turions break off and sink to the bottom where they spend the winter. Next spring, they float to the surface and grow into new plants. Alongside Duckweed at about one millimeter in length, is the world's smallest flowering plant, Northern Watermeal, which has no stems or roots and floats on the surface.

OTHER PLANTS | BUR-REED

Simple-stemmed Bur-reed

Bur-reeds, including Simple-stemmed Bur-reeds, are named for their spiky, ball-shaped flower and fruit clusters.

Great Bur-reed

Great Bur-reed bears the largest ball-shaped clusters of its group. Like many other aquatic plants, a Bur-reed's stem contains canals called aerenchyma that supply air to the submersed root-like rhizomes (horizontal stems).

OTHER PLANTS | BUR-REED

Floating Bur-reed

The strap-like leaves of Floating Bur-reed usually have a much greater physical presence than the flowers themselves.

Narrow-leaved Cattail

A native to Eurasia, the Narrow-leaved Cattail is an invasive species in many regions. The leaves are narrower than those of Broad-leaved Cattails, and there is a gap between the lower female and upper male flower spikes. However, this species regularly hybridizes with the native Broad-leaved Cattail.

OTHER PLANTS | CATTAIL

Broad-leaved Cattail

The dense flower spikes of the native Broad-leaved or Common Cattail are arranged with the male flower spike touching the female one situated below it. Also known as bulrushes, cattails reproduce by cloning as well as by seed, and thus can quickly dominate a wetland.

LICHENS | **OTHER PLANTS**

Crustose Lichens

The original graffiti artists, lichens paint rocks in rainbow hues. Lichens are a symbiotic combination of a fungus and either a photosynthesizing alga or cyanobacterium, or both. Crustose Lichens are a huge group that form crusts on rocks, and, along with foliose and fruticose lichens, cover granite so that eventually its original color — pink — is no longer visible.

OTHER PLANTS | LICHENS

Elegant Sunburst Lichen

1 One of the most beautiful foliose (leaf-like) lichens, Elegant Sunburst Lichen likes calcium, so it lives on rocks with at least some of that chemical present. It is incredibly hardy; most of the specimens subjected for one and a half years to the stresses of space outside the International Space Station survived the ordeal!

2 This unusual pattern is due to Elegant Sunburst Lichen growing where Common Terns "enriched" the island on which they nested.

Gray Reindeer Lichen

Gray Reindeer Lichens are found on northern talus slopes and rocky outcrops. As the name suggests, they are important winter food for Woodland Caribou. Fur traders made tea with Gray Reindeer Lichen but in northern Europe and Russia, along with other lichens, it was used in the production of brandy.

OTHER PLANTS | LICHENS

Star-tipped Reindeer Lichen

Star-tipped Reindeer Lichen has a yellowish-greenish tinge and forms round-topped mounds. It is the lichen commonly used in floral decorations and in architectural and railroad models.

LICHENS | **OTHER PLANTS**

Arctic Frosted Rock Tripe

Arctic Frosted Rock Tripe is an Umbilicaria lichen, a group named after how they attach to a rock face by one central contact structure (like an umbilical cord). Rock Tripes have been used as food by indigenous peoples around the world, and have also served as emergency food, most notably for members of the ill-fated Franklin expedition.

Appendix

Reference Books

All of the following are excellent resources, and most of them are portable. I recommend the most recent edition. If the volume is out of print, a used copy or library loan might be available.

The Peterson Field Guide series
National Geographic Field Guide to the Birds of North America — Dunn and Alderfer
The Sibley Guide to Birds — Sibley
Field Guide to the Dragonflies and Damselflies of Algonquin Park and Surrounding Area — Jones et al
Dragonflies and Damselflies of the East — Poulson
Damselflies of the Northeast — Lamb
The Mayfly Guide — Caucci
Caddisflies — Ames
Spiders of the North Woods — Weber
Spiders of North America — Rose
Ferns and their Allies of the North Woods — Walewski
Northeast Ferns — Chadde
Sedges of Maine — Arsenault et al.
Sedges of the Northern Forest — Jenkins
Grasses of the Northern Forest — Jenkins and Engstrom
Newcomb's Wildflower Guide — Newcomb
Shrubs of Ontario — Soper and Heimburger (out of print)
Trees in Canada — Farrar
Lichens : the Macrolichens of Ontario and the Great Lakes Region of the U.S. — McMullin

Online Resources

Canadian Journal of Arthropod Identification
https://cjai.biologicalsurvey.ca
This is a superb free on-line resource with visual keys to many groups of insects and other arthropods. There are plenty of excellent photographs for the species and groups covered.

Bugguide https://bugguide.net
Bugguide is an excellent online resource devoted to North American insects, spiders and their kin, offering identification, images, and information.

eBird https://www.ebird.org/
eBird is an incredible resource for reporting and keeping records of all your bird sightings during your travels anywhere in the world, and seeing where other birds have been seen.

iNaturalist https://www.inaturalist.org
iNaturalist is a first-rate resource in which you can post photos of plants or animals for confirmation of your ID, or to acquire one. Observations are added to a data base so that every species has its location documented, and a map allows you to see where sightings were made. Plus, there are plenty of photos to look at!

Go Botany https://gobotany.nativeplanttrust.org/
Go Botany is an exceptional resource for plant identification, offering keys to all groups and diagnostic photos for each species covered.

Index

Alder Flycatcher, 55
American Bittern, 49
American Bittersweet, 313
American Black Duck, 15
American Bladdernut, 296
American Common Reed, 342
American Coot, 47
American Cow-parsnip, 226
American Crow, 37
American Mink, 87
American Mountain-ash, 274
American Nursery Web Spider, 183
American Pipit, 65
American Redstart, 81
American Robin, 273, 282
American Toad, 107
Aquatic Leaf Beetle, 172
Arctic Frosted Rock Tripe, 355
Aurora Damsel, 122
Autumn Meadowhawk, 148

Backswimmer, 174
Bald Eagle, 29
Balsam Fir, 265
Balsam Poplar, 277
Baltimore Checkerspot caterpillar, 223
Bank Swallow, 39
Barn Swallow, 41
Barred Owl, 67, 275
Baskettail, 140–141
Basswood, 275
Bayonet Rush, 328
Bearded Sedge, 330
Beaver, 98, 187, 188, 278, 305
Beaver Lodge, 190
Bebb's Willow, 305
Beck's Beggarticks, 213
Beck's Water-marigold, 213
Belted Kingfisher, 30
Birds, 7–83
Black Ash, 272
Black-backed Woodpecker, 69
Black Bear, 89, 282
Black-capped Chickadees, 258
Black Chokeberry, 285
Black Fly, 15
Black-shouldered Spinyleg, 133
Black Spruce, 267
Black Tern, 27
Black Vulture, 32
Blanding's Turtle, 115
Blue Dasher, 144
Blue Flag, 242
Blue-headed Vireo, 76
Blue Vervain, 246
Blue-winged Teal, 13
Blunt Spikerush, 335
Bog Aster, 243
Bog Bean, 209
Bog Birch, 304
Bog Laurel, 299, 300
Bog Rosemary, 300
Bog Willow, 307
Boneset, 225
Boreal Bluet, 124
Boreal Snaketail, 136
Bristly Rose, 301
Broad-leaved Cattail, 349, 350
Broad-leaved Meadowsweet, 297
Broad-winged Hawk, 34
Brown Drake, 168
Buckbean, 209
Bulblet-bearing Water-hemlock, 227
Bullfrog, 102
Bullhead Pond-lily, 212
Bumble Bee, 243
Bur-reeds, 346–347
Buttonbush, 291

Cackling Goose, 8
Caddisflies, 164
Canada Bluejoint, 340
Canada Darner, 130
Canada Goose, 8, 191
Canada Manna Grass, 341
Canada Rush, 327
Cardinal Flower, 235
Carex Sedge, 329
Caspian Tern, 26
Cattails, 349–350
Cedar Waxwing, 78, 273, 282
Chalk-fronted Corporal, 153, 239
Cherry-faced Meadowhawk, 147
Chestnut-sided Warbler, 83
Choke Cherry, 281, 282
Cinnamon Fern, 319
Clam Sign, 194
Clubtail, 135–139
Common Arrowhead, 205
Common Basketail, 140
Common Bladderwort, 214–215
Common Gallinule, 46
Common Goldeneye, 18
Common Grackle, 58
Common Green Darner, 128
Common Loon, 22
Common Merganser, 20
Common Nighthawk, 31
Common Raven, 38, 198
Common Tern, 27, 352
Common Whitetail, 150
Common Yellowthroat, 59
Cottonwood, 280
Crimson-ringed Whiteface, 289
Crustose Lichens, 351

Damselflies, 118–127
Dark Fishing Spider, 182
Deer Fly, 160
Dock, 234

357

Dot-tailed Whiteface, 145
Double-crested Cormorant, 23
Downy Woodpecker, 70
Dragonflies, 118–154
Dragonfly Nymph Case, 196
Dragonhunter, 134
Dwarf Birch, 304

Eastern Amberwing, 142
Eastern Chipmunk, 94, 282
Eastern Coyote, 91
Eastern Gatersnake, 110
Eastern Hemlock, 268
Eastern Kingbird, 73
Eastern Phoebe, 75
Eastern Pondhawk, 143
Eastern Ribbonsnake, 111
Eastern White Cedar, 258–259
Eastern White Pine, 260–261
Eastern Wolf, 90
Ebony Jewelwing, 119
Elegant Sunburst Lichen, 352
Emerald Ash Borer, 272
European Common Reed, 342
European Frog-bit, 206

Fall Sneezeweed, 254
Fawn Darner, 132
Feeding Platform, 189
Female Melittid bee, 250

Ferns, 316–326
Flat-topped Aster, 231
Flies, 155–163
Floating Bur-reed, 348
Flowering Rush, 220
Flowerless Stalks, 188
Four-spotted Skimmer, 154
Fragrant Wood Fern, 323
Freshwater Sponge, 178
Fringed Sedge, 333
Frogs, 102–107
Frosted Whiteface, 146

Giant Eastern Crane Fly, 155
Giant Mayfly, 169
Glossy Buckthorn, 309
Golden Eagle, 29
Grasses, 340–343
Grass-leaved Goldenrod, 253
Grass Pink, 233
Gray Catbird, 77
Gray Dogwood, 288
Gray Reindeer Lichen, 353
Gray Squirrel, 96
Great Blue Heron, 48, 51
Great Bur-reed, 347
Great Egret, 52
Greater Duckweed, 345
Greater Yellowlegs, 64
Great Horned Owl, 66
Green Alder, 303
Greenfrog, 103
Green Heron, 53
Green-winged Teal, 16

Hairy Woodpecker, 70
Herptiles, 100–115
Herring Gull, 25
Highbush Cranberry, 295
Hooded Merganser, 19
Horned Bladderwort, 251
Horse Fly, 161

Insect Feeding, 192
Insects, 116–176
Interrupted Fern, 319

Jack Pine, 263
Jewelweed, 255

Killdeer, 60

Labrador Tea, 290
Lake Darner, 129
Lakeside Sedge, 329
Lancet Clubtail, 138
Larch *see* Tamarack
Least Bittern, 50
Leatherleaf, 289
Leech, 180
Lesser Bladder Sedge, 332
Lichens, 351–355
Lilypad Clubtail, 135
Little Floating-heart, 210
Little Stout Crawler Mayfly, 167
Long-beaked Willow, 305
Long-horned Caddisfly, 166
Long-legged Fly, 158

Maidenhair Spleenwort, 326
Mallard, 14
Mammals, 84–99
Map Turtle, 114
Marginal Wood Fern, 324, 325
Marsh Bellflower, 224
Marsh Cinquefoil, 217
Marsh Fern, 320
Marsh Skullcap, 248
Marsh Wren, 54
Mink, 193, 198
Mink Frog, 104
Moose, 93, 305
Moose Fly, 162
Mosquito, 157
Mountain Holly, 292
Muskrat, 99
Muskrat Lodge, 191
Mute Swan, 9

Narrow-leaved Cattail, 349
Narrow-leaved Meadowsweet, 297
Ninebark, 286
Nodding Beggartick, 252
Nodding Ladies' Tresses, 221
Northeastern Sedge, 334
Northern Cardinal, 312
Northern Flicker, 71
Northern Harrier, 42
Northern Leopard Frog, 105
Northern Shoveler, 13
Northern Spreadwing, 127

Northern Water-plantain, 204
Northern Watersnake, 108–109
Northern Wild Rice, 343

Olive-sided Flycatcher, 74
Orange Bluet, 126
Oriental Bittersweet, 313
Osprey, 28
Ostrich Fern, 321

Painted Turtle, 112, 181
Peregrine Falcon, 36
Phantom Crane Fly, 156
Pickerel Frog, 106
Pickerelweed, 245
Pied-billed Grebe, 21
Pileated Woodpecker, 72, 312
Pin Cherry, 281
Pine Siskin, 258
Pitcher-plant, 236
Polymorphic Pondweed Moth, 175
Pondweed, 344
Poplar *see* Trembling Aspen
Porcupine, 97, 275
Powdered Dancer, 121
Prickly Wild Rose, 301
Prince Basketball, 141
Purple Loosestrife, 219, 245
Purple-stemmed Aster, 244

Raccoon, 88, 198, 275
Raven Sign, 198
Red Crossbill, 260–261

Red-eyed Vireo, 76
Red Fox, 195
Red Maple, 270
Red-osier Dogwood, 287
Red Pine, 262
Red-shouldered Hawk, 33
Red Squirrel, 95, 258
Red-tailed Hawk, 35
Red-winged Blackbird, 57
Reed Grass, 340
Ring-billed Gull, 24, 179
Ring-necked Duck, 17
Riverbank Grape, 312
River Horsetail, 316
River Jewelwing, 118
River Otter, 86, 186, 189, 193
Rock Polypody, 325
Rose Pogonia, 233
Rosy Meadowsweet, 298
Round-leaved Sundew, 239
Royal Fern, 318
Ruby Meadowhawk, 147
Ruby-throated Hummingbird, 235
Ruffed Grouse, 278
Rush, 327
Rusty Cliff Fern, 322
Rusty Crayfish, 179
Rusty Snaketail, 137
Rusty Woodsia, 322

Sandhill Crane, 48
Scarlet Tanager, 269
Scat, 193, 195
Scent Mound, 187

Sedges, 327–339
Sensitive Fern, 317
Seven-angled Pipewort, 207
Shadfly, 168
Shadow Darner, 131
Sheep Laurel, 299, 300
Shining Willow, 306
Short-eared Owl, 43
Showy Mountain-ash, 273, 274
Shrubby Cinquefoil, 283
Shrubs, 283–309
Silver Maple, 271
Simple-stemmed Bur-reed, 346
Skimmer, 151–152, 154
Skimming Bluet, 123
Slaty Skimmer, 152
Small Purple Fringed Orchid, 241
Smooth Serviceberry, 284
Smooth Turtle Leech, 181
Snakes, 108–111
Snapping Turtle, 113, 181
Solitary Sandpiper, 63
Song Sparrow, 79
Sora, 45
Speckled Alder, 302, 303
Sphinx Ladies' Tresses, 221
Spoon-leaved Sundew, 238
Spotted Joe-Pye Weed, 240
Spotted Sandpiper, 62
Spotted Water-hemlock, 228

Spring Azure caterpillar, 297
Square-stemmed Monkeyflower, 247
Stable Fly, 163
Star-tipped Reindeer Lichen, 354
Steeplebush, 298
Stonefly, 170, 196
Stream Bluet, 125
Stream Cruiser, 139
Sugar Maple, 269, 270
Swamp Aster, 244
Swamp Birch, 304
Swamp Candle, 249
Swamp Loosestrife, 219, 245
Swamp Milkweed, 237
Swamp Rose, 301
Swamp Sparrow, 56
Swaying Bulrush, 338
Sweetflag, 211
Sweet Gale, 308
Swift River Cruiser, 139

Tall Cottongrass, 337
Tall Meadow-rue, 230
Tall White Aster, 231
Tall White Bog Orchid, 222
Tamarack, 264
Tawny Cottongrass, 337
Track, 197
Trees, 258–282
Tree Swallow, 40
Trembling Aspen, 278–279
Trumpet-net Caddisfly, 165
Trumpeter Swan, 10

Tube Maker Caddisfly, 197
Tufted Yellow Loosestrife, 250
Tule Bluet, 124
Tundra Swan, 11
Turkey Vulture, 29, 32
Turtle Nest, 195
Turtles, 112–115
Tussock Cottongrass, 336
Tussock Sedge, 331
Twelve-spotted Skimmer, 149

Umbilicaria lichen, 355

Variegated Pond-lily, 212
Vines, 310–313
Violet Dancer, 120
Virgin's Bower, 310
Virginia Creeper, 311
Virginia Rail, 44

Water Boatmen, 174
Water Horsetail, 316
Water-parsnip, 229
Water Strider, 173
Water Willow, 219, 245
Whirligig Beetle, 171
White Birch, 276
White-faced Meadowhawk, 147

White Meadowsweet, 297
White Panicled Aster, 232
White Spruce, 266
White-tailed Deer, 92
White-throated Sparrow, 80
White Turtlehead, 223
White Water-lily, 202
Widow Skimmer, 151
Wild Calla, 203
Wildflowers, 200–256
Wild Iris, 242
Wild Raisin, 294
Willow Flycatcher, 55

Wilson's Snipe, 61
Winterberry, 292, 293
Wood Duck, 12
Woodland Caribou, 353
Woodsia, 322
Woody Plants, 256–314
Woolgrass, 339

Yellow-bellied Sapsucker, 68
Yellow Loosestrife, 249
Yellow Pond-lily, 212
Yellow Warbler, 82

Zebra Caddisfly, 165

Dedication

For Britta, who for years has paddled her own canoe but now paddles with me as one.

Acknowledgments

A special thanks to Bill Crins, Stephen Darbyshire and Don Sutherland, for sharing their unrivaled expertise, and Willow Reed, for technical assistance. And to my Britta, for steadying our canoe during those frequent stops for photographs.